THE GEORGIAN BAY SHIP CANAL

CANADA'S ABANDONED NATIONAL DREAM

RAY LOVE

FriesenPress

Suite 300 - 990 Fort St
Victoria, BC, V8V 3K2
Canada

www.friesenpress.com

Copyright © 2021 by Ray Love
First Edition — 2021

All rights reserved.

No part of this publication may be reproduced in any form, or by any means, electronic or mechanical, including photocopying, recording, or any information browsing, storage, or retrieval system, without permission in writing from FriesenPress.

Photographers: Jessica McShane, Terence Hayes.

ISBN
978-1-03-910499-0 (Hardcover)
978-1-03-910498-3 (Paperback)
978-1-03-910500-3 (eBook)

1. HISTORY, CANADA, POST-CONFEDERATION (1867 TO PRESENT)

Distributed to the trade by The Ingram Book Company

CONTENTS

1 The Route	4
2 The Geology of the Route	7
3 Early History of the Route: Indigenous Peoples	11
4 Arrival of Europeans	15
5 Early Canals in Upper Canada	31
6 Treadwell's Treatise	37
7 Shanly and Clarke Surveys	41
8 The Competition	50
9 The Lobbyist	57
10 The 1904 to 1908 Public Works Survey	64
11 Laurier and the Railway Mess	82
12 The Georgian Bay Canal versus the Welland Canal	90
13 The French River Improvement and Power Scheme	97
14 The Sifton Plan	102
15 Power Plays	111
16 Bill No. 78	120
17 The Railway Commission Hearing	125
18 Conclusion	129
Epilogue	131
Author's Note	135
Acknowledgements	140
Endnotes	142
List of Illustrations	150
Bibliography	154

This book is dedicated to Bob and Susie Perry and their 'Little bit of Heaven' on the French River.

INTRODUCTION

"If you live to be as old as I am now you will see a double track around Lake Superior together with a ship canal by the Ottawa."[1]

—Sir John A. Macdonald, first prime minister of Canada

"I am perfectly satisfied that the Ottawa Valley presents the greatest facilities of any route upon the continent for the transportation of the products of the North-West to the Atlantic Ocean."[2]

—Alexander Mackenzie, second prime minister of Canada

"Canada's chief item of expenditure is in creating routes for her commerce. She has to complete her railway system and build the Georgian Bay Canal."[3]

—Sir Wilfrid Laurier, seventh prime minister of Canada

"Both the Welland and the Georgian Bay Canal are worthy of consideration by the government, and will receive that consideration; and I am sure if we come to the conclusion that both are advisable, the resources of Canada will be sufficient to carry out both undertakings."[4]

—Sir Robert Borden, eighth prime minister of Canada

"Had such a route existed in the United States it would have been canalized years ago."[5]

—New York Engineering News

Despite the views expressed by these prominent individuals and organizations, the Georgian Bay Ship Canal was never built. The route was surveyed over a dozen times and contracts were let for its construction, but little work was actually completed.

The idea of a canal following the route of the voyageurs captured the imagination of Canadians for centuries. One author has aptly named it the "Trans-Canada Highway No. 1." [6] In the political sphere, it was discussed and debated numerous times in the Parliaments of the combined Canadas and the Dominion of Canada. These debates lasted two years shy of a century before the idea was put to rest on May 1, 1927 by a vote in the House of Commons.

This canal scheme was one of numerous ideas put forward to create a shipping lane to join the largest reservoir of fresh water on the planet, the Great Lakes, to the Atlantic Ocean. Although the Great Lakes are essentially one large river flowing from Lake Superior to the Atlantic, there were several obstacles to a southerly shipping route. The two most glaring of these were Niagara Falls and the Lachine Rapids. The northern route from Georgian Bay along the French River and Lake Nipissing, down the Mattawa and Ottawa Rivers to the St. Lawrence was seen as a legitimate alternative.

A parallel was envisioned between these routes and the European transportation network which connected the Black and Mediterranean Seas with the Atlantic Ocean. The ultimate prize was ease of access to mineral resources and grains from the interior of the North American continent.

At various times, the idea was courted by British industrialists who seemed more than eager to build it. Boards of Trade in cities such as Montreal, Ottawa, North Bay, and Port Arthur sent delegations of up to 3,500 citizens to lobby for it. The who's who of Canadian politics and business from the early 1800s to 1927 had a say in its promotion.

Yet it was never built. Was it a bad idea? No, it was commendable in many ways. Was it a failure? No, more a missed opportunity for Canadian national and international financial success.

In one sense the demise of the project was a blessing for the ecological health of the waterways involved. Little did early shipping interests know of the possibility of invasive species from foreign seas wreaking havoc with the Great Lakes ecosystem. The term ecosystem did not grace the pages of a dictionary until coined in by Oxford ecologist Arthur Tansley in 1935.

The canal was a true enigma, as in times of unbridled enthusiasm for new public works in Canada, it never saw the bucket of a steam shovel. What follows is a detailed examination of various attempts to build the canal, its promoters, and its challengers.

1
THE ROUTE

The canal has had many names over the years, including the Montreal and Lake Huron Canal, the Champlain Ship Canal, the Ottawa Waterway, the Montreal, Ottawa and Georgian Bay Canal, and simply the Georgian Bay Ship Canal. For the purposes of this book, we will use all of the above and also refer to it simply as "the route."

The best description of the route of the proposed canal comes from a Scottish geographer by the name of Martin C. Comrie. His information came from an exhaustive four-year survey completed between 1904 and 1908 by the Canadian Department of Public Works at a cost to Canadian taxpayers of slightly less than seven hundred thousand dollars.

> *The proposed Georgian Bay Ship Canal is essentially a river and lake canalisation scheme and would utilise natural waterways, which fortunately exist almost in a continuous line from Georgian Bay on Lake Huron to Montreal, the most inland and most important of the Canadian ocean ports. A straight line drawn through Montreal and Sault St. Marie has a direction almost due east and west, and follows closely the Ottawa River and Lake Nipissing, which thus furnish the most direct and shortest route from Lake Superior to a seaport. This route, if it can be made navigable for large lake freighters, appears to be the natural outlet for all the commerce of the West seeking transport through Lakes Superior and Michigan to the nearest ocean port. The distance between Georgian Bay and Montreal by the proposed route is 440 miles, of which from 410 to 420 miles follows the course of some river or lake.*

From Georgian Bay to the water-parting between the Ottawa River and the Great Lakes, a distance of 81 miles, the canal would follow the channels formed by the Pickerel and French Rivers, and Lake Nipissing. The French River system enters Georgian Bay by three openings. For 6 miles the waterway would be coincident with the middle channel, on which the French River Village is situated. Then, after passing through the main channel of the French River, the canal would proceed by way of that part of the French River waters known as the Pickerel River as far as the east side of Cantin Island. A sharp turn north would then be made to rejoin the main channel of the French River, which would direct the route as far as Lake Nipissing.

From Lake Nipissing, through the highest level on the route, it is proposed to construct an artificial waterway for 3 miles, with the exception of stretches covered by a few small lakes. This artificial channel would lead into Trout Lake, thence into Turtle Lake, the Little Mattawa River, and Talon Lake, which would be used as far as Sand Bay at the eastern end, a distance altogether of 21 miles. From Sand Bay there would be a canal for 3 miles to the Mattawa River, which would be utilised as far as the town of Mattawa, a distance of 13 miles, where another canal-cut, three quarters of a mile in length, would make an entrance into the Ottawa River. This river, which expands into large and deep lakes in many places, would be followed all the way down to the foot of the Lake of Two Mountains, a distance of 293 miles. From the foot of the Lake of Two Mountains to Montreal, a distance of 25 miles, either the St. Lawrence River or a branch of the Ottawa River, called Rivière des Prairies, flowing north of the Island of Montreal, may be utilised. The former route would involve 5 miles of artificial waterway, and the latter about 11 miles. By the first route the canal would enter Montreal Harbour at its upper end, by the second the St. Lawrence ship channel would be joined at Bout de l'Île, some 11 miles below the eastern boundary of Montreal Harbour, or 17 miles below the city customhouse. For the whole route the aggregate length of purely artificial waterways would be astonishingly small, being estimated at 28 miles. Apart from these, about 80 miles of lake and river beds would require to be improved by dredging or excavation, leaving 332 miles of natural waterways, wider than 300 feet, and over 22 feet in depth, not requiring any improvement. At the outlet of the French River is French River Village harbour. The river has an average width of 500 feet, with a depth of about 30 feet of water. Outside, the Bustard Islands, about 3 miles distant from the mouth of the harbour, offer good protection against southerly winds, which are sometimes of considerable violence. [7]

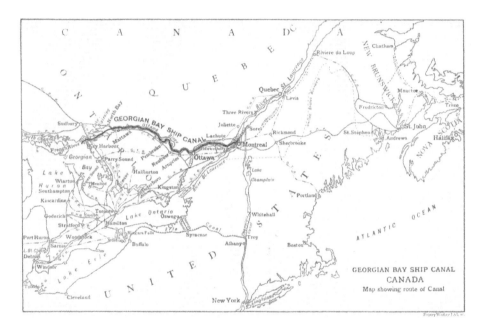

2
THE GEOLOGY OF THE ROUTE

The route flows primarily through the Canadian Shield, a Precambrian rock formation that makes up almost half of Canada. It is home to some of the earth's oldest rock, which has been altered by the forces of heat and pressure many times in its three-billion-year history. This area has been lifted up by tectonic forces and lowered by various forms of erosion to create the highlands we see today. It is also home to a rift valley called the Ottawa and Bonnechere Graben, which runs between two faults lines, the Mattawa and Petawawa faults. These faults create a topographical depression running from Montreal to Lake Nipissing, some 700 kilometres. On the western end of the rift valley, the faults are fractured into smaller fault lines running from Lake Nipissing to Georgian Bay.

Most of the region is made up of hard rock, either igneous rock such as granite, quartz, and diorite, or metamorphic rock, for example, schists, gneiss, and basalt. The eroded sediments of these highlands have been deposited farther south to form another geographic region, the Great Lakes-St. Lawrence Lowlands. This area is composed of sedimentary rock. The southern end of the Ottawa River flows through this much softer rock of limestone, sandstone, and shale.

The retreat of the Wisconsin glacier, the last to come and go across this region of Canada, created the river basins and lakes that make up the route.

"By 12,000 years ago, the ice front had retreated to an east-west line running along the north side of the present Ottawa Valley, but it still blocked the great lake named Algonquin, that had been formed by glacial meltwaters over rather more than the area now occupied by Lake Huron and Georgian Bay.

Far to the east the recession of ice had permitted the earlier Atlantic Ocean to come flooding up the St. Lawrence Lowlands as the Champlain Sea." [8]

The glacial Lake Algonquin broke through the land barrier and drained east and then south into the Champlain Sea through the Ottawa Valley. As the glaciers melted, they created huge volumes of water filled with rock debris that acted like a giant scraping tool carving out the rock over which they flowed. This created many unique rock outcrops, some of which can be seen in the following photos.

Glacial scouring on the French River

As the massive ice sheet left the area, the land underneath began to rebound. Approximately 4,000 years ago, the French River reversed its course and began to flow west from Lake Nipissing into Georgian Bay.

The two drainage basins are now separated at the summit, which is a relatively insignificant height of land about four miles east of the city of North Bay.

The route can be roughly divided into the Nipissing and French River system flowing west and the Mattawa and Ottawa Rivers flowing east, then south. These two branches and the small lakes that separate them create a waterway of 689 kilometres.

The French River west of Highway 69

The characteristics of each water body vary throughout the route based in part on the geology of the various sections. A fuller description of the hydrologic conditions of the route are to be found later in the book.

3

EARLY HISTORY OF THE ROUTE: INDIGENOUS PEOPLES

THE ROUTE OF THE PROPOSED SHIP CANAL CLOSELY FOLLOWED THE INDIGENOUS TRADE routes, established many centuries before the arrival of Europeans. At the time of European contact, the three Indigenous groups who inhabited the area encompassing the route were the Algonquin, Nipissing, and Ottawa, all tribes speaking a version of the Algonquin dialect. All three tribes were Ojibwa in heritage, today calling themselves Anishinabek, or "the people."

The Ottawa were noticed by Champlain picking and drying blueberries at the mouth of the French River. They also inhabited Manitoulin Island and ranged west into the current states of Michigan and Wisconsin. The Nipissing inhabited the north shore of Lake Nipissing, the Sturgeon River, Trout Lake, and parts of the French River. The Algonquin had a headquarters on Allumette Island in the Ottawa River near present day Pembroke and could be found throughout the Ottawa Valley. All three groups were hunter-gatherers that subsisted on hunting, fishing, trapping, collecting berries, and growing a bit of corn. They all used common technology such as birch-bark canoes, snowshoes, toboggans, and birch-bark bowls.

Even prior to contact with the French, the tribes were active traders. As an example, the Nipissing traded as far west as Lake Nipigon, north to Hudson's Bay, south into Huron Territory in southern Georgian Bay, and east into Quebec. They traded dried pickerel, considered a delicacy, and furs for corn, tobacco, and fishing nets, amongst other products. They also traded for flint and red ochre. Some of the furs traded made their way to Tadoussac at the mouth of the Saguenay River. Here there was a large Indigenous fur trading market.

Trade in furs and foodstuffs were common amongst many tribes. The Hurons were active trading partners adept at growing and preserving a variety of vegetables such as corn, beans, and squash. The Nipissing were strong allies of the Hurons for a time, so much so that some members would winter near the Hurons in order to gain access to a supply of maize or corn.

The Nipissing (Nbisiing) were prominent around the French River area and left their mark in the form of three rock paintings, two near the confluence with Lake Nipissing and one above Récollet Falls, closer to Georgian Bay. The Algonquin tribes were a part of the larger Ojibwa nation that had a very well developed division of labour. The men hunted and trapped and made snowshoe frames and toboggans; the women fished, cooked, tanned leather hides, and made clothing for the family. "Thus the Ojibwa phrase an objectively cooperative economy in the most individualistic terms. The man hunts alone on his isolated trails, the wife works alone in the wigwam … and they exchange the products of their work." [9]

The name Nipissing (Nbisiing) means "people of the little water," which is the Indigenous description for Lake Nipissing as compared with Lake Huron. According to the *Jesuit Relations* they were considered sorcerers due to the abundance of medicine men in their ranks and their rituals involving gymnastics and juggling. "Neighbouring tribes were so convinced that the Nipissing could inflict evil at remote distances that they crossed their territory with trepidation and loaded with gifts of appeasement." [10] The Nipissing also regularly gathered at a multiday Feast of the Dead, which involved digging up the graves of those who had passed away since last year's feast. Described in the *Jesuit Relations*, the ceremony involved the exchange of gifts between tribes, singing, dancing, and athletic competitions. At the feast were held the elections of Nipissing chiefs.

> *This election was followed by the Resurrection of those Persons of importance who had died since the last Feast; which means that, in accordance with the custom of the Country, their names were transferred to some of their relatives, so as to perpetuate their memory. On the following day, the Women were occupied in fitting up, in a superb manner, a Cabin with*

an arched roof, about a hundred paces long, the width and height of which were in proportion... After that, the same Women carried the bones of the Dead into this magnificent Room. These bones were enclosed in caskets of bark, covered with new robes of Beaver skins, and enriched with collars and scarves of porcelain beads. Near each dead body sat the women, in two lines, facing each other. Then entered the Captains, who acted as Stewards and carried the dishes containing food. This feast was for the Women only, because they evince a deeper feeling of mourning. Afterward, about a dozen Men with carefully selected voices entered the middle of the Cabin, and began to sing a most lugubrious chant, which, being seconded by the Women in the refrains, was very sweet and sad. [11]

In the spring and summer, most families of Nipissing would settle along the French River, Georgian Bay, or Lake Nipissing, making maple syrup in the spring and harvesting berries in summer. In the autumn, fish would be speared and dried for winter. During the winter, the families would disperse through the woods to hunt moose and other game.

The Algonquin people can be traced back eight thousand years in occupying the Ottawa Valley. The introduction of European diseases and wars with the Iroquois permanently changed their status as an important player in the fur trade. This, coupled with being ignored by the English in the handing out of reserves, changed the way of life of the Algonquin forever. They did name the site of present day Mattawa as the "Meeting of the Waters" and had a permanent settlement there until the mid-nineteenth century. A reservation for the Algonquin at Golden Lake was established in 1873 and remains the sole Algonquin reserve today. Presently the Algonquin are spread amongst ten communities in North-Central Ontario. Community members continue to work toward a land settlement of the former Algonquin territory.

The Nipissing (Nbisiing) on Reservation No. 10 on the outskirts of North Bay fared somewhat better than the Algonquin. Many fled to Lake Nipigon during the Iroquois Wars, only to return four years later to resume trapping and trading in furs. Their reserve, located on the shores of Lake Nipissing and encompassing 51,910 acres, is the home to some 918 band members living locally as well as 1,939 residing off-reserve. Their members are involved in a variety of local businesses, including a commercial fishery on Lake Nipissing.

Also set aside by the crown was Reservation No. 13, which is closer to Georgian Bay in the Parry Sound District. It is occupied by the Amikwa, or Beaver People who are also Anishinabek and part of the Henvey Inlet First Nation. At this time, this group is seeking more information on their history.

4

ARRIVAL OF EUROPEANS

Beginning in the early 1600s, the route was travelled by French explorers, missionaries, and fur traders beginning with Etienne Brûlé in the fall of 1610. He was accompanied by Algonquin guides. The French and later the English relied on Indigenous groups in their pursuit of furs. They showed traders how to build canoes, fish, and travel in winter using snowshoes, toboggans, and dog teams. They were of great assistance in showing Europeans where and how to obtain furs.

Samuel de Champlain ventured to Huronia along the route in 1615 in order to cement a healthy business relationship with the Hurons. After his visit, the Huron fur canoes regularly travelled the route, bringing tens of thousands of beaver skins to French traders at Quebec and Montreal. The Hurons were successful in establishing the fur trade with the French at the expense of their northern Algonquin neighbours. The Nipissing and Ottawa did serve for a time as middlemen in the fur trade, taking European goods such as rifles, knives, cooking pots, and blankets to more remote tribes for barter and then exchanging the furs so obtained with the French.

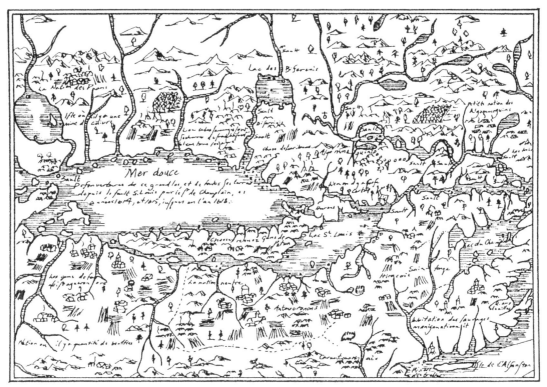

CHAMPLAIN'S MAP OF 1632.

Champlain was preceded in his first voyage by an anxious Recollét priest, Joseph Le Caron, who wanted desperately to be the first missionary to bring the Christian god to the natives. He was followed by others of his order such as Gabriel Sagard, who in imitation of Saint Francis, went through the woods of the Canadian Shield barefoot. Sagard writes of his travels:

"It would be hard to tell you how tired I was with paddling all day, with all my strength, among the Indians; wading the rivers a hundred times or more, through the mud and over the sharp rocks that cut my feet; carrying the canoes and luggage through the woods to avoid the rapids and frightful cataracts; and half-starved all the while, for we had nothing to eat but a little *sagamite*, a sort of porridge of water and pounded maize, of which they gave us a small allowance every morning and night." [12]

Plaque at the French River Visitor Centre

The "black robes," or Jesuits, followed the Recolléts and established a permanent fortified mission on the banks of the Wye River, Saint-Marie-Among-the-Hurons, near present-day Midland, Ontario. Huron furs would be transported from Ste. Marie up Georgian Bay to the mouth of the French River, then inland via the route. The Jesuit fathers and their French workmen established this settlement at Huronia but brought with them contagious diseases. In 1634, disease swept through the Hurons, decimating the tribal population by half.

Recollét Falls

The settlement at Huronia met its demise at the hands of the warring Iroquois, who attacked and annihilated the settlement in 1648 and 1649, killing Hurons and Jesuit priests, including the torturing to death of Fathers Brebéuf and Lalemant. In the spring of 1649, the Jesuits burned the settlement and escaped to Christian Island on Georgian Bay. The Iroquois persisted in their disruption of the fur trade of other nations, attacking canoe convoys on the route for the next several decades. Their actions caused the dispersal of the Nipissing to the Lake Nipigon area and the Algonquin to the Lac St. Jean region and to Jesuit missions at Quebec and Montreal. The Nipissing eventually returned to their homeland with the peace treaty between the French and Iroquois in 1667. The Algonquin were largely decimated by disease.

The fur trade was so lucrative that independent white traders such as Médard Chouart Des Groseilliers and his brother-in-law, Pierre-Esprit Radisson, began to venture into fur-trading country and buying furs directly from Indigenous peoples rather than through native middlemen. These supreme individualists, termed *couriers des bois*, began living with Indigenous peoples to learn their ways as well as to buy furs. It is estimated that there were five hundred of these independent businessmen operating in the Lake Superior region by 1680. Radisson and Des Groseilliers, on one trading expedition in 1660, returned to Montreal along the route carrying a fortune in furs packed into sixty canoes paddled by three hundred natives. At the same time, the British were exploiting the furs in Rupert's Land with the Hudson Bay Company, establishing trade with Indigenous tribes along the rivers that flowed into Hudson and James Bay.

The French in Quebec continued a successful fur trade until the 1750s, when war with the English broke out in the colonies. The war was lost to the English in 1760 at the Plains of Abraham, and the fur trade subsequently took on a new direction.

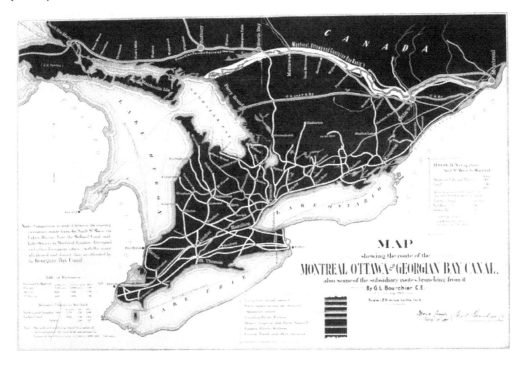

The new fur trade entrepreneurs were Scottish, English, and American businessmen headquartered in Montreal. After several tries at going it alone, they formed a loose collection of traders called the North West Company. Principal amongst these men were the Scots Simon McTavish and the great explorer Alexander Mackenzie. These men sought new and better sources of furs in what are now the Prairie Provinces and the Northwest Territories, establishing a supply chain that stretched from Montreal to Fort William and again to Fort Chipewyan on Lake Athabasca. The volume of furs they harvested was astonishing: "…in the summer of 1798 were delivered 106,000 beaver skins, 32,000 martin skins, 1,800 mink skins and 40,000 other skins of a dozen varieties."[13] The bulk of these skins were transported to Montreal via the Ottawa waterway.

Mackenzie's travels in search of additional sources of furs between 1789 and 1793 took him first to the Arctic Ocean and back along the river that now bears his name. The second voyage to the Pacific Ocean cemented his fame as one of the greatest explorers of all time.

In order to transport furs such long distances, two types of canoes were developed. Both still used the birch-bark technology of the Indigenous peoples. The *canot du maître*, or Montreal canoe, was designed for lake travel. It was approximately thirty-six feet long by five feet wide and drew three feet of water. It had a capacity of 7,700 pounds (3,492 kg) and was paddled by eight to twelve men. It was used primarily to carry furs from Fort William to Montreal via the route. The smaller *canot du nord*, or north canoe, twenty-five feet in length, was used on rivers to the west and northwest of Lake Superior.

These canoes were paddled by the famous *voyageurs*, mostly French Canadian men but also natives, blacks and Scotsmen. These men regularly paddled fourteen hours per day and were responsible for carrying the cargo along portages, each man carrying 180 pounds at a time. They split up the distances by carrying for approximately one third of a mile, putting the packs down, resting, and then resuming the portage.

David Thompson, one of Canada's early travellers and arguably the best early cartographer in Canada, said of the voyageur, "When he has a moment's respite he smokes his pipe, his constant companion and all goes well; he will go through hardships but requires a full belly at least once a day, good tobacco to smoke, a warm blanket and a kind Master who will take his share of hard times and be first in danger." [14]

The length of a canoe journey was often measured in "pipes." These were the number of smoke breaks allowed in paddling between portages or across a large lake.

The voyagers had to meet certain physical requirements to obtain employment. Author Grace Lee Nute writes that they were small men, no larger than five foot six or 165 centimetres, so as not to take up too much cargo space. "Though lacking in stature, the voyageur had to be incredibly strong ... This meant that the typical voyageur was a man with mammoth upper body and tiny legs."[15] The life of a voyageur was tempered by adventure; as one man put it, he had saved ten lives, had twelve wives and six dogs, and spent all his money on pleasure.

Each year, the voyageurs began their journey up the Ottawa in early May. Their canoes were loaded to the gunnels with small farm animals, religious paraphernalia, tools, wine and spirits, cutlery, earthenware, writing tools and paper, medicines, food stores, cutlery, kitchenware, blankets, rolls of cloth, weapons, and other trade goods. Then came the paddlers and passengers. These floating hotels first stopped at St. Anne de Bellevue church, some two miles from the western extremity of the island of Montreal. There they made a donation to have a priest say a small mass that was to ensure their protection on the upcoming voyage.

The next stop on the route for the voyageurs was at the upper end of the Lake of Two Mountains, where after the first day's paddle, each canoe received a keg of rum. No doubt this was a popular stop.

From Hawkesbury to Ottawa was an uneventful sixty miles, but from Hull to present-day Aylmer, Quebec, were a series of rapids at Chaudière, which required either the use of poles or rope lines to advance the canoes upstream. This process was termed lining or tracking.

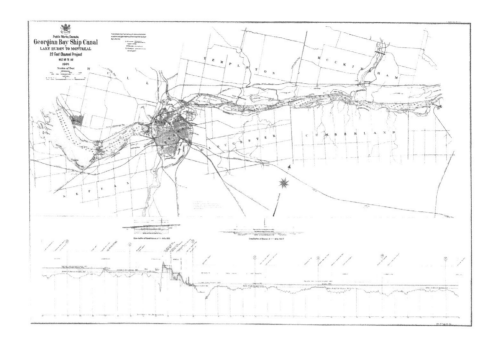

Map of Ottawa from 1908 survey

No. 7. --- Parliament Hill and Rideau Canal Locks at Ottawa.

Ottawa in 1908 and today

The next obstacle was picturesque Chats Falls, where the Precambrian Shield and its harder, more erosion-resistant rock, was encountered. Rapids again were forged at Portage-du-Fort and the Allumette Islands. At Allumette, the Ottawa was split into four channels by islands, and the voyageurs had to pick the channel of least resistance. Further upstream at Deep River, the Canadian Shield once again exerted its influence on the landscape with a 500-foot-high granite wall and unfathomably deep water. Pointe-de-Baptime at Deep River was where the rookie voyageurs were ceremoniously baptized in the river, naked regardless of the water temperature.

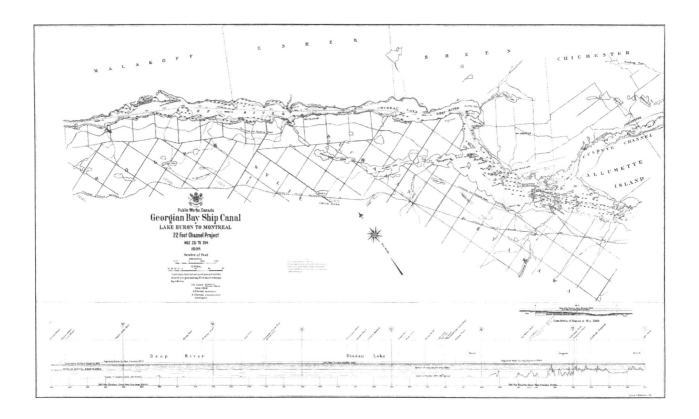

Photo and Map of Deep River

The last significant portage was at Rapides-des-Joachims prior to turning onto the Mattawa River. The confluence of the Mattawa and Ottawa River was marked by three crosses erected in 1686 by order of French Captain Sieur de Troyes to commemorate where Champlain had stopped to repair his canoe in 1615. The crosses, which have been used as a landmark to locate the river, have been repaired and replaced until present time. The Mattawa River was a rocky and rugged stretch of forty miles with eleven portages. At the end of the Mattawa, another ceremony was held as the voyageurs threw away their poles since they were soon to be travelling downstream with the current. Trout Lake was the head of the Mattawa, and just west of it was the divide separating the Ottawa and St. Lawrence River watersheds. The divide was an unimpressive swampy area of some seven miles in length, five of which could be paddled.

No. 14. --- Confluence of Mattawa and Ottawa Rivers.

The confluence today

Lake Nipissing was an easy paddle on a calm day, and another sixty-five kilometres brought the voyageurs to its western terminus at yet another set of rapids named Chaudière.

The seventy miles past this point were on the French River, which amounted to a single day's paddle with two short portages, all of it with the current. The voyageurs had three outlets of the French River into Georgian Bay to choose from. It is generally believed they chose the Western Outlet, which differed from the plans for the canal.

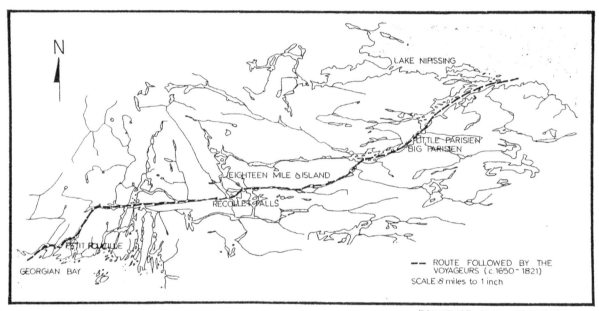

ROUTE OF THE VOYAGEURS

The voyageurs were a colourful lot that kept themselves occupied in racing against the other canoes and in singing as they paddled. Each voyageur was said to know fifty songs. The songs, such as "Alouette" and "À la Claire fontaine," remain cultural treasures of Canada. The singing unquestionably helped to allay the extreme physical demands of the job.

"Of such use is singing, in enabling the men to work eighteen and nineteen hours a day (at a pinch). Through forests and across great bays, that a good singer has additional pay. The songs are sung with might and may, at the top of the voice, timed to the paddle, which makes about fifty strokes a minute." [16]

The job of the voyageur was at times dangerous, as evidenced by a memory from David Thompson: "They preferred to run the Dalles (rapids on the French River); they had not gone far, when to avoid the ridge of waves, which they ought to have kept, they took the apparent smooth water, were drawn into a whirlpool, which wheeled them around into its vortex; the canoe with the men clinging to it, went down end foremost, and all were drowned; at the foot of the Dalles search was made for their bodies, but only one man was found, his body much mangled by the waves." [17]

Typically, the men that were recovered were placed in shallow graves marked by wooden crosses. When voyageurs came across these gravesites in their travels, their caps were doffed and a prayer uttered.

Competition for increasingly scarce furs between the North West Company and Hudson Bay Company led to their amalgamation in 1821 under the name of the former. The famous governor of the HBC, George Simpson, kept the fur trade alive from 1821 to 1860. Dubbed the "Little Emperor," Simpson was small in stature but large in presence. He kept himself in fine physical condition by paddling as well as taking a morning swim in whatever body of water he was near, regardless of the time of year. He was always immaculately dressed, as were his paddlers, and always brought a piper with him on his journeys. He ran the Hudson's Bay Company with hands-on efficiency, travelling each year to visit as many outposts as time warranted. He made the trip along the route twenty-nine times to visit various interior forts from the company headquarters in Montreal. Simpson also left us accounts of the lives of the voyageurs, which form an integral part of Canadian history.

> *Such was the routine of our journey, the day, generally speaking, being divided into six hours of rest and eighteen of labour. This almost incredible toil the voyageurs bore without murmur, and generally with such a hilarity of spirit as few other men could sustain in a single forenoon. But the quantity of work, even more decidedly than the quality, requires operatives of iron mould. In smooth water, the paddle is applied with twice the rapidity of the oar, taxing both lungs and arms to the greatest extent; amid shallows, the canoe is literally dragged by men, wading to their knees or their loins ... in rapids the towing line has to be hauled over rocks and stumps, through swamps and thickets ... Of the baggage, each man has to carry at least two pieces, estimated at one hundred and eighty pounds*

avoirdupois, which he suspends in slings of leather across the forehead, so that he has his hands free to clear the way… [18]

Simpson's successor as governor of the HBC was Alexander Grant Dallas, who along with his secretary, Edward Hopkins, travelled from Fort William to Montreal via the route to assume his post. Hopkins' second wife, Frances Anne Hopkins (daughter of British hydrographer and rear admiral William Beechley), was to immortalize the voyageurs with her paintings. Often accompanying her husband on business trips for the company, Frances made numerous sketches and paintings of scenes along the route, as well as of the men and their canoes that guided the Hudson Bay parties. She initially disguised her sex by signing her paintings FAH, since women artists were seldom recognized. She was, however, able to exhibit her paintings at the Art Association of Montreal and on numerous occasions at the Royal Academy in London. Two of her voyageur paintings are shown here.

Canoe manned by Voyageurs passing a waterfall, by Frances Anne Hopkins

Shooting the rapids, by Frances Anne Hopkins

Beginning in the 1840s, cheaper silk hats were in fashion in Europe, and the fur trade gradually declined. After three hundred years, the reign of the mighty Canadian beaver had ended. The outcome was that a nation had been in large part defined by the fur-trading routes that stretched from Atlantic to Pacific. It fell to politicians to make the new country a reality.

The history of Indigenous tribes along the route was greatly impacted by the Robinson-Huron Treaty of the early 1850s. In it, they surrendered the title to the territory between Southern Georgian Bay and Sault St. Marie extending inland to the height of land north of Lake Huron, as well as all lands not ceded within Canada West. In exchange, they were given an annuity, reserves pertinent to this study on Lake Nipissing (No. 10) and on the French River (No.'s 9 and 13), as well as rights to hunt, fish, and trap on ceded lands.

A signatory to the treaty was Michel Dokis, who remained chief of his small band of Nipissing for fifty-six years until his death in 1906. He negotiated the treaty on behalf of his band and received Reservation No. 9, a parcel of land where Lake Nipissing empties into the French River. Much of the area is an island named Ockickandawt and contains rapids at Big and Little Chaudière.

Chief Dokis was remarkable in many ways. At the treaty hearings, he pledged never to sell the timber from his reserve to the Crown or any lumber company and held true to his word. Despite many attempts to sway him to sell the million dollars' worth of timber, he would not waver from his promise.

In 1856, the HBC became interested in the cranberries that grew near the HBC post on the Sturgeon River. They wished to pick and sell them. The chief intervened by written letter and effectively petitioned for the land to be left to natives who had picked berries in the marsh since time immemorial.

He then convinced agents of the Crown that they had mistaken the size of his reserve on the French River. He had it extended from nine square miles to 60.99 square miles. It was eventually expanded again to its present size of 39,790 acres. The chief and his family did not initially live on their reserve but were granted a prime piece of land on Lake Nipissing Reservation No. 10 named Dokis Point by Nipissing Chief Shabogishic. It was from there that he conducted his business as a fur trader. The chief did take up his reserve on the French River in 1895. In 1897, the reserve numbered seventy-five residents.

In order to secure furs from Lake Temagami, Chief Dokis opened a fur-trading post next to the Hudson's Bay Company Post in 1864 and sent two of his sons to manage it. They were hugely successful due to their understanding of native languages and their intermarriage with native women from the area.

Clearly, the chief and his family had gained over the years the respect of agents of the Crown. "One was apt to consider them rather as members of a commercial firm in good standing."[19]

In 1904, the federal Department of Public Works was studying the possibility of building a canal from Lake Nipissing to Georgian Bay. This would involve a 1,400-foot (427 metres) canal and lift lock at the Big Chaudière Falls. At no time was Chief Dokis consulted, even though the canal ran through his reserve. The *Indian Act* of 1876 took for granted that reservation lands could be appropriated by any level of government wanting a way through it.

5
EARLY CANALS IN UPPER CANADA

THE FIRST CANALS CONSTRUCTED ALONG THE ROUTE WERE CREATED IN DEFENSE OF BRITISH North America. After the bloody spat between Britain and the United States named the War of 1812, the security of the colonies was uppermost in the minds of British defense authorities. In particular, the supply line between Montreal and Kingston was thought vulnerable. An American general divulged after the war had ended that his forces had a plan in place to destroy this link. An alternate route between Montreal and Kingston was sought using the Ottawa and Rideau rivers, the Rideau Lakes, and the Cataraqui River. This bypass began with the canals on the Ottawa River, the Carillon and Grenville canals, which were begun in 1819 and completed in 1834. The building of the Rideau Canal began in the fall of 1826 and was completed in 1832. The Rideau Canal, with forty-seven locks and more than fifty dams, was viewed as one of the engineering marvels of its day. These canals opened up traffic first for Durham boats (large wooden freight boats) and then for steam-propelled vessels.

Dates and Dimensions, Canadian Locks

YEAR		LENGTH Ft.	WIDTH Ft.	DEPTH Ft.
1708	Sault Ste. Marie	38	8¾	2 on sill
1780	Locks at Cascades and Coteau	35	6	2½
1804	Locks at Cascades and Coteau	110	20	4
1819	Military canals, Ottawa River (Grenville)	106½	19¼	6½
1825	Lachine Canal	100	24	4½
1829	First Welland Canal (wooden locks)	110	22	8
1832	Rideau Canal	134	33	5
1834	Grenville Canal (Ottawa River)	130¾	32½	6½
1834	Carillon Canal (Ottawa River)	126½	32	6
1834	Chute-a-Blondeau (Ottawa River)	131	33	6
1843	Ste. Anne Lock (Ottawa river)	190	45	6
1843	Chambly Canal (Richelieu River)	118	55	9
1843	Cornwall Canal (St. Lawrence River)	200	55	9
1846	Beauharnois Canal (St. Lawrence River)	200	45	9
1846	Second Welland Canal	150	26½	10¼
1847	St. Ours Lock (Richelieu River)	200	45	7
1880	Culbute (Ottawa River) wooden lock	200	45	4
1890	St. Lawrence and Welland	270	45	14
1890	Grenville (Carillon and Ste. Anne)	200	45	9
1892	Sault Ste. Marie	900	60	19

The building of canals and the many proposed builds were an attempt by Canada to outflank the Americans in developing an all-Canadian waterway. This was planned to create a Montreal based trading network that would challenge the Erie Canal-Hudson River route to the port of New York. Unfortunately, the Erie Canal was completed in 1825 and siphoned off most of the western trade. The Ottawa Waterway was but one such supposed rival to "Clinton's Ditch," the local name for the Erie Canal named after the Governor of New York. Unfettered by the early success of the Erie Canal, a report to the Legislative Assembly of Ontario urged action: "It is between Canada and the State of New York that the struggle for the carrying trade of the Western Country will be fought, and if Canada does not display the greatest possible activity, she will have to succumb to her southern rival. It is true that national pride and immense capital and the beaten track of commerce are on the side of New York; but God and Nature are stronger than all these, and let any man compare the 'Erie Ditch' with the mighty St. Lawrence…"[20]

In the nineteenth century the main tributary of the St. Lawrence, the Ottawa River, was an important waterway to the settlers who were developing farms along its shores. The settlers needed supplies brought in to carve out their forest homesteads. Initially a combination of carts, Durham boats and freight canoes were used to get supplies from Montreal to towns such as Hawkesbury and Grenville. These were replaced as early as 1822 by steam powered vessels. During the remainder of the century, steamships plied the Ottawa between the major waterfalls and rapids and transported passengers, mail and freight along the river. It was possible for a number of years for passengers to sail between Ste. Anne near Montreal all the way to Mattawa on large well-appointed steamships that got smaller as the journey continued. At each falls or set of rapids, such as Chats Falls, Allumette Rapids and Des Joachims, passengers would disembark onto carriages for the ride to the next vessel upstream. The entire trip took three days. One can envisage the combination of luxury vessels, spectacular scenery and overnight stays at riverside hotels being an incredibly charming experience.

The Ottawa Valley was known for its vast timber reserves. In particular majestic stands of white pine destined for navies around the world were common. The lumber industry began in the early 1800's and dominated activity along the river for the remainder of the century. Logs were cut and squared by honed craftsmen and organized into cribs which in turn were lashed together into large rafts to be floated down the Ottawa. The raftsmen, of which there were up to 13,000 by mid-century, had their own primitive accommodation on board the raft. They were responsible for navigating rapids, locks, canals, timber slides, etc. to get their logs to market.

As early as 1831 sawmills using water power sprung up along the river and its tributaries. Companies were able to saw their logs into more manageable size for transport. These smaller products were typically called slabs. With the advent of steam driven sawmills the process became more efficient. In 1896 some 343 million board feet of lumber was being produced by mills in the Ottawa area alone. Barge building ensued at Hull and these barges, towed by steam powered tugs, carried up to 300,000 board feet of lumber each. The destinations varied from travel down the Ottawa to Montreal, to Kingston via the Rideau Canal and New York via the Erie Canal. An active timber trade on the Ottawa River continued until World War 1.

The Mattawa River was also an important source of white pine and was a lumbering hub throughout the nineteenth century. It boasted its own Paul Bunyan like legend in Big Joe Mufferaw. Originally named Joseph Montferrand, he was a French Canadian logger, strongman and folk hero to working men. There are many tales of his incredible feats of strength and endurance some of which live on in the song "Big Joe Mufferaw", written by Canadian folk singer 'Stompin' Tom Connors. According to Tom, Big Joe "paddled up the Ottawa all the way to Mattawa in just one day."

In central Ontario, the Trent Valley Canal connecting Trenton, Ontario with Georgian Bay was surveyed as a military route. The route went from the mouth of the Trent River on the Bay of Quinte north to Rice Lake, north-west through the Kawartha Lakes to Lake Simcoe, and out the Severn River to Georgian Bay. The first canal was built in 1833 but was determined to need too many locks to be feasible for military operations. Several locks were built, but work was halted after the 1837 Rebellion and not taken up again until the 1880s.

Another route, the Scugog route, was to run from Whitby on Lake Ontario through to Lake Scugog, then Lake Simcoe and on to Georgian Bay. This scheme lasted about as long as it took to introduce it to the Legislative Assembly of Upper Canada. Another proposal was the Chatham Route from Lake Erie through to Lake Huron; however, it did not save much distance or time from the existing Welland-St.

Lawrence route. Yet another scheme was to build a canal from Burlington Bay to Lake Huron, but standing in the way of this dream was the massive Niagara Escarpment.

Toronto investors were lured to consider the Georgian Bay Ship Canal, similar in name to the route, which proposed to build a canal from the Humber River through to Lake Simcoe then via the Nottawasaga River to Georgian Bay. This would eliminate the need to pass through Lake St. Clair, the St. Clair and Detroit Rivers, as well as Lake Erie and the Welland Canal. This route had strong backing from the city of Toronto and sought federal government support. The one serious disadvantage of this canal was the Oak Ridges Moraine, a large and deep glacial deposit that would have to be cut through in order to build a canal. The cuttings in some areas approached two hundred feet of depth.

The plans were initially developed by architect Kivas Tully and later modified by American engineer Henry Spalding. The Toronto and Georgian Bay Canal Company was formed in 1856, and various investigations were made into its feasibility. The scheme was altered in 1894 to include hydro power developed along the route and sold to the city of Toronto. The new company was named the Georgian Bay Ship Canal and Power Aqueduct Company. Company members were unable to convince the city fathers of the feasibility of the scheme, and the company charter lapsed in 1896.

A hybrid development was the ship railway between the head of navigation on the Humber River to the Nottawasaga River and hence to Georgian Bay. The rail line would only be a mere sixty-six miles in length, with hydraulic lifts at both ends and a reinforced three-track rail line. A ship of up to 2,000 tons could be loaded onto a rail car and pulled the distance by five locomotives. The backers of this proposal, were not unable to convince the government in power of its feasibility.

Ship railways were common in other parts of the world. In Canada, a ship railway was begun in 1888 on the Chignecto Peninsula bordering Nova Scotia and New Brunswick. Unfortunately for many, the Chignecto line, financed by Baring Brothers and Company of England, became short of funds, and the project was halted in 1891 after being seventy-five percent completed. Despite repeated attempts, the federal government refused to put up monies to complete the railway. It is believed that the failure of the Chignecto Railway was one reason Laurier was reluctant to support the Georgian Bay Ship Canal.

GENESIS OF THE GEORGIAN BAY CANAL CONCEPT

The idea put forth by British colonizers of being able to sail between Montreal and Georgian Bay had its early genesis as a defense measure, since a military base had been established by the British at Penetanguishene on Georgian Bay. The British Navy wished to circumvent the water bodies shared with the United States including the St. Lawrence River and Lakes Ontario and Erie. In 1819, young officers of the Royal Engineers travelled up the Ottawa and through what is now Algonquin Park, searching for a route. The idea was promoted again in 1829 by Colonel John By, who had gleaned it from an Ottawa River settler, Charles Shirriff. Shirriff's sons would become surveyors and would survey a portion of the route during their careers.

In 1835, the government of Upper Canada sent Lieutenants Carthew and Baddeley to survey the route and determine if the area was suitable for settlement. The response was somewhat pessimistic. Baddeley returned to the area in 1836–37 accompanied by one of Canada's great explorers and mapmakers, David Thompson. Thompson was sixty-six years of age at the time, broke due to investments gone bad, ailing from a bad hip, and blind in one eye. Regardless, the map work he completed along this journey was superb. His maps were used by the Canadian government as official maps of the country for the next century.

In 1839, Thomas Hawkins surveyed the western portion of the route from Georgian Bay to Lake Nipissing. He believed there was a suitable harbour at French River Village and a navigable waterway along the French River with ample water depth and only three sets of rapids. His main concern was the depth of Lake Nipissing and the availability of water at the summit to fill ship locks. This was followed in 1845 by a survey by William Logan of the Geologic Survey of Canada of the Upper Ottawa, the Mattawa, and the summit area near Lake Nipissing.

These early surveys pointed to the route being a viable option, since it was a remarkable natural waterway. Citizens and businessmen along the Ottawa River began to talk of the commercial possibilities of the route and to lobby governments to take action. For the next eighty-two years, a battle was waged between supporters of the canal, critics, and various levels of government.

6

TREADWELL'S TREATISE

CHARLES P. TREADWELL, OTTAWA VALLEY NATIVE AND SHERIFF OF THE COMBINED COUNTIES of Prescott and Russell, put together a book entitled *Arguments in favor of the Ottawa and Georgian Bay Ship Canal*, written in 1855 and published in 1856. It was a compilation of letters to politicians, newspaper articles, reports from captains of steamships plying the Ottawa River and findings of river surveys. It was produced to lobby the government of the combined Canadas to build the canal. Clearly, it was intended to improve the business fortunes of the citizens of Ottawa, who were in competition with Toronto for commercial and industrial supremacy in emerging Ontario.

In the book, it states that Lower Canada and the Ottawa Valley were being ignored by governments in the railway-building craze in British North America in favour of the southern portion of Upper Canada. In 1846, canals had been built on the St. Lawrence and at Welland, which allowed through traffic from the Upper Lakes to the St. Lawrence. Most of this traffic, Treadwell argued, even that carried by Canadian ships, was being funnelled through Buffalo via railway or along the Erie Canal to New York. This had hurt the shipping business along the Ottawa River and at Montreal. The residents of these cities wanted their own waterway to attract shipping from the upper Great Lakes.

This began a series of arguments in favour of the route, which would be repeated in various forms, added to, and refined for the next seventy-five years. The most salient point was the distance saved along the Georgian Bay route. Chicago, the chief transshipment point for products of the American midwest to Montreal, Canada's chief port, was 1,680 miles (2,703 kilometres) along the St. Lawrence route. The distance was shortened to 971 (1,562 km.) along the Georgian Bay Canal. The distance from Chicago to Liverpool, Britain's main port for ocean-going vessels, was 4,683 miles (7535 km.) via the St. Lawrence versus 3,630 (5482 km.) via Georgian Bay. The calculations of the day indicated a saving on shipping of fifty-four cents per bushel of wheat using the Georgian Bay route. If freight was diverted at Buffalo to sail from New York, the distance and cost to Liverpool was increased yet again.

A legitimate and often overlooked argument in favour of canals was that they were a much cheaper form of bulk transport than their nearest competitor, the railways. Estimates vary over the decades, but figures are from three to nine times less expensive. The shipping season was limited in North America to approximately seven months, but during those months, water transport was the least expensive means of shipping bulk goods.

The Montreal Board of Trade commissioned its own survey in 1855, which was published in Treadwell's book:[21]

> *The Ottawa Canal — A canal to connect the Georgian Bay with the St. Lawrence River has long been a favourite project for opening communication between the upper lakes and the ocean. This was one of the plans submitted to the late Ship Canal Convention, and a report just published from a committee of the Montreal Board of Trade gives the result of special inquiries into this subject. This committee recommend the route already surveyed — from the mouths of the French River, on the Georgian Bay, by way of Lake Nippisinque and the Matawan and Ottawa Rivers to Montreal. The distance from Chicago to Montreal, by proposed route, is allowed to be only 980 miles, against 1,348 by the Welland Canal, or 808 miles shorter than the existing route. The lake navigation (including Nippisinque) would be 575 miles; river 347; canal, 58. The entire expenditure required, with locks 250 feet long by fifty feet wide, and a depth of ten feet throughout, is estimated at $14,000,000. Immense advantages are anticipated from the increase of manufacturing along the route, the opening of a new market for the lumber of the Ottawa region, the enhanced value of the mineral deposits so lavishly scattered through that portion of the province, while from*

the copper districts of Lakes Huron and Superior the trade of Quebec and Montreal would derive large profits.

Treadwell and others writing in his book bolster their case by pointing out that the Georgian Bay route was a "red" or all-Canadian route in the instance of armed conflict with the United States. The threat of such a conflict was ever present in pre-Confederation Canada, especially at the conclusion of the American Civil War.

In terms of defense, the site of Ottawa was seen as having a strong military position on a hill overlooking the river. The idea occurred to the Duke of Wellington, former British general and twice British prime minister, that if Ottawa were combined with the city of Hull, Canada's two founding European nations could be fused into the capital of the two Canadas. This was recommended, and Ottawa became the functional capital of the Canadas in 1866 and the capital of the Dominion of Canada one year later.

Treadwell's book continues with the arguments that the economic development that would occur with the building of the canal would be significant in cities such as North Bay, Ottawa, and Montreal. The canal would also provide impetus for the settlement of the Ottawa Huron Tract, which had not been settled by Europeans and would not be to any great degree for decades to come. The agricultural potential of the French River area, which was a key to settlement, was somewhat suspect, as Champlain's quote indicates.

"Having rested two days with the chief of the said Nipissings, we re-embarked in our canoes and entered a river flowing out of this lake, and made some thirty-five leagues along it, and passed several little rapids, some by portaging, others by running them, as far as Lake Attigouautan (Georgian Bay). This whole region is even more unprepossessing as the former, for I did not see in the whole length of it ten acres of arable land…"[22]

There were other resources to be exploited! Those resources were primarily the rich timberlands of the Ottawa valley and what is presently Algonquin Park. One estimate stated that there were one hundred and fifty years' worth of timber to be harvested.

In the treatise were interviews with steamship captains who worked on the Ottawa River. At the time, there was a regular passenger service between Montreal and Ottawa. The sailors believed the river to have promise as a ship canal due to its width and large volume. A rather optimistic reporter for the *Ottawa Citizen* noted in the conclusion to the book that "no power on earth can retard or prevent the accomplishment of the work."

TABLE OF DISTANCES IN STATUTE MILES OF WATER ROUTES.

PROPOSED NEW CANADIAN ROUTE.

Via Great Lakes, Georgian Bay Ship Canal and Montreal.	Distance to Montreal.	Distance Montreal to Liverpool via Belle Ile.	Total Distance.
Fort William to Liverpool	934	3,189	4,123
Duluth "	1,056	3,189	4,245
Milwaukee "	906	3,189	4,095
Chicago "	972	3,189	4,161

PRESENT CANADIAN ROUTE.

Via Great Lakes, Welland and River St. Lawrence Canals and Montreal.	Distance to Montreal.	Distance Montreal to Liverpool via Belle Ile.	Total Distance.
Fort William to Liverpool	1,216	3,189	4,405
Duluth "	1,338	3,189	4,527
Milwaukee "	1,176	3,189	4,365
Chicago "	1,242	3,189	4,431

UNITED STATES ROUTE.

Via Great Lakes, Erie Canal, Hudson River and New York.	Distance to New York.	Distance New York to Liverpool.	Total Distance.
Fort William to Liverpool	1,358	3,571	4,929
Duluth "	1,480	3,571	5,051
Milwaukee "	1,318	3,571	4,889
Chicago "	1,384	3,571	4,955

7
SHANLY AND CLARKE SURVEYS

THE PROTESTATIONS OF BUSINESSMEN IN MONTREAL AND OTTAWA DID NOT GO UNHEEDED, as in 1856, the government of the Province of Ontario commissioned another survey. Again, the purpose of this survey was to gather information on the feasibility of a canal.

The man the government of Ontario chose was civil engineer Walter Shanly. He was born in the Republic of Ireland and moved with his family to the London, Ontario area in 1837. He and his brother Francis both apprenticed as engineers. Walter cut his engineering teeth on locks and canals working on the Saint-Ours lock on the Richelieu River, the Beauharnois Canal, and the Welland Canal. He and Francis turned their skill to railway building mid-century and assisted in building the Bytown and Prescott Railway and the Toronto and Guelph Railway. Walter held the position of general manager and chief engineer on the Grand Trunk Railway, which at the time was the longest railway in the world. Later in his career, he combined engineering work with politics, becoming an MLA in 1863 and MP in 1867. Walter remained active as an engineer and consultant into his eighties.

Shanly began his survey with two parties, one that he headed from the Mattawa River to Lake Nipissing and another along the Ottawa River from Deep River to Chats Lake (Lac des Chats). He began in August of 1856 but was called off the job by the governor general in May of 1857. His second crew continued work until the project was terminated in January of 1858. Despite the cancellation, Shanly and crew had paddled much of the route, and he was able to produce a report that outlined its character. In his own words:

"I voyaged the whole of the above mentioned portion of the route, some 260 miles, by canoe… and I reached my journey strongly impressed with the conviction that nature had thus marked out a pathway in the desert that the Genius of Commerce will, at no far off day, render subservient to its end; the navigable connection of the Great Lakes with 'La Grande Rivière du Nord,' I look upon as inevitable."[23]

Shanly noted that trade, especially in grain, was growing in the Western United States, and he predicted a similar increase in Canada when Rupert's Land was settled. He viewed the route as a series of large rivers and lakes requiring few canals. He did tip his cap to the St. Lawrence route as being easier to construct due to the nature of the soil and rock needed to be excavated, but he also recognized the shipping distance and cost advantage of the Georgian Bay route. He was able to find a suitable entrance to the French River and a location for a port, protected from Georgian Bay by the Bustard Islands. This was the result of having an experienced native guide. He was quite proud of this finding, since it had previously been missed by the famous British naval cartographer, Captain Bayfield.

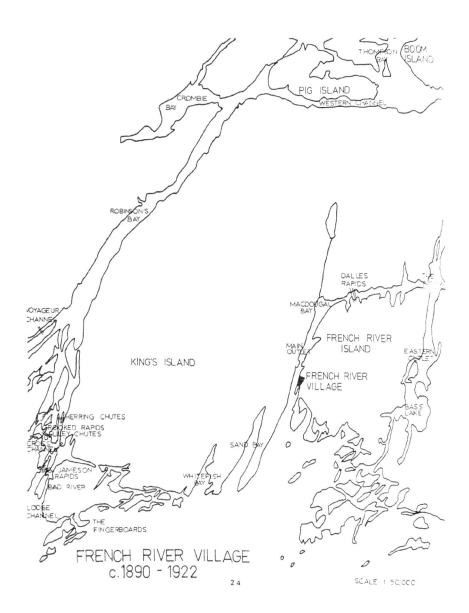

Entry to the French River via the middle or main outlet

The French River he found to be more a series of long deep lakes within high walls and having deep water. The few rapids he thought would be easily surmountable. Likewise, the connection from the French River to Lake Nipissing was easy to construct, and Lake Nipissing itself offered thirty miles (50 km.) of lake shipping.

French River looking toward Lake Nipissing

At the summit, about four miles (6.5 km.) east of Lake Nipissing, between it and Trout Lake, Shanly envisioned a problem. He initially could not fathom how enough water could be stored in the area to provide for locks and canals along the route. His solution was to dam several outlets of Lake Nipissing, raising the level of the lake some twenty-three feet to the level of Trout Lake. In 1856, this would flood some arable land, but since there were few inhabitants, it would essentially not cost the government a great deal in expropriation fees.

Trout Lake, Turtle Lake, and Talon Lake led to the Mattawa River, the largest tributary of the Ottawa. Once the Ottawa River was reached, long expanses of wide, deep water were the norm. Certainly, there were rapids to be surmounted, especially in shield country, but Walter envisioned that these could easily be overcome.

THE GEORGIAN BAY SHIP CANAL

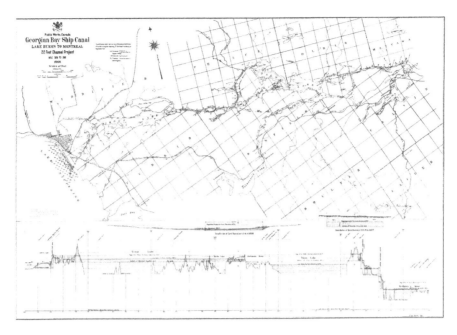

Summit section of the route

In his summation, Shanly saw advantages to the route, as well as challenges. The advantages not already stated included the possibility of tourism along this scenic route, the availability of wood to power steamships, and the opening of the area to settlement. He envisioned Ottawa becoming an industrial centre not only for wood manufacturing but also for flour milling. He recognized the vast potential of water power along the Ottawa. He also saw the advantages to Montreal as a trans-shipment point between the smaller inland lake boats and ocean-going vessels. New Yorkers were at the time considering building the Caughnawaga Canal, which would run from the St. Lawrence east of Montreal down Lake Champlain to join with the Hudson River. This would provide a combined waterway from Chicago to New York.

The challenges, he believed, were fewer than the advantages. The hard rock of the Precambrian Shield was an issue for excavation and for building locks. The solution was to bring in limestone for lock construction from Manitoulin Island. An additional issue was that there were no roads or rail lines in the area to get supplies and workers to job sites. The bourgeoning railway system in the Province of Canada would soon take care of this concern.

Shanly's estimate of the cost for the entire route was $24 million, which was steep for the times. His solution was to start small and build one section at a time. Walter's plan was shelved shortly after completion, probably due to the enormous cost. He would continue to advocate for the route as a Canadian MP after Confederation. Meanwhile, a cheaper estimate emerged.

Thomas Curtis Clarke, an American by birth, was commissioned to survey the route yet again in 1859. Clarke is known in Canada as the engineer who built the East and West Houses of Parliament in Ottawa. He came up with similar conclusions as Shanly on the economic benefits of the Georgian Bay Canal plan. He reworked the engineering at the summit as to only raise the level of Lake Nipissing by fifteen feet, thus drowning less arable land around the lake's perimeter. He and fellow engineer E.R. Blackwell then proceeded to examine the three outlets of the French River into Georgian Bay. The most surprising feature of his report was the costing, coming in at $12 million to Shanly's $24 million. Clarke justified his numbers by relying on more dams to raise water levels and the building of fewer canals.

There were now three reports in the sunset years of the 1850s for the Government of the Province of Canada to consider. In what was to become a typical response, the government of the day did nothing to implement any of the suggested works. An explanation for inaction was written in the *Ottawa Gazette*. The canal was not begun "principally for the reason that the greater number of the leading men of the province are personally interested in the improvement of the St. Lawrence (route)."[24]

Shanly's report was delivered to the Legislative Assembly in 1858 and Clarke's in 1860, but both were called back to testify in an inquiry held in the spring of 1863 as to the feasibility of the route. The inquiry was chaired by Ottawa politician, surveyor, and journalist, Robert Bell.

In the Bell inquiry, engineers Shanly, Clarke, and Gallway were interviewed, as was an Ottawa River boat captain Sclater. The members of the inquiry also interviewed a number of Ottawa and Montreal merchants.

Seven questions were put to each participant. In twenty-first century language they were:

1. Describe the suitability of the route for propeller driven steamships.
2. What are the advantages of this route over the St. Lawrence route and the Toronto to Georgian Bay Canal?
3. What are the current trade possibilities, and how would they change with the building of the route?
4. What are the physical characteristics of the area through which the route passes?
5. What would be the effect on commerce and settlement on building the canal?
6. Is the route advantageous for the military security of Canada?
7. How would you propose to construct such a route?

A number of interesting points came out of the deputations. Shanly and Clarke wrestled with explaining the discrepancy of $12 million in their quotes but in the end managed to come within $6 million of one another. The increase in the grain trade out of Chicago was noted as 12,863,912 tons in 1854 versus 56,477,104 tons in 1862. This was all American grain, since the Canadian West had yet to be put to the plough. Ira Gould, an American businessman working in Montreal, claimed that one half of all American grain exports could be funnelled through the Georgian Bay route. Captain Sclater made an important observation in that grain would keep better in the bowels of ships along the route due to colder air and water temperatures. Another excellent point was made by businessman Robert Eisdale. At the time, Chicago was booming and buying a great deal of its construction lumber from Canadian firms. This would, in his estimation, provide the perfect return cargo for ships along the route.

In terms of opening up the area through which the route passed, it was noted that there were vast timber resources to be developed, as well as water power to provide for lumber mills. A few areas of arable soil were pinpointed along the South River and Lake Nipissing in Upper Canada and the valley of the Blanche and Lake Timiskaming in Lower Canada. In general, the Ottawa Valley was seen as suitable for agricultural settlement, the Mattawa and French River areas less so.

In his final report, Robert Bell trusted the reports of the engineers which indicated that the route could be managed with current technology. The Ottawa route was seen as the shortest and the cheapest of the three routes examined. The trade opportunities were viewed as tremendous and growing annually. The timber and mineral resources along the route were noted, as was the possibility of industrial development due to abundant water power. The military advantages of an all-red route inland from the United States were obvious. Finally, in the discussion of how to build the route, private enterprise was the first choice. The government could assist with providing a guarantee of interest on the monies raised by company bonds. Robert Bell then introduced a bill in 1864 to incorporate a canal company consisting of forty-nine individuals with the purpose of building the canal. It was time to build!

It was becoming evident to the businessmen of Ottawa, however, that strong opposition was developing amongst those that favoured the St. Lawrence route. In a book on Canadian canals published in 1865, the opinion was ventured that the Ottawa route would not improve the commercial relations of the province as a whole and was injurious to the west. It was regarded in no other light than a local improvement.

This statement flew in the face of both fact and attitude, since the western provinces of Canada were strong supporters of the Georgian Bay Ship Canal route. Further discussion of the route would have to wait until the Confederation of Prince Edward Island, New Brunswick, Nova Scotia, Quebec, and Ontario into the Dominion of Canada.

The work of Shanly and Clarke was used in the 1870's by the Montreal Northern Colonization Railway as justification for a rail link between Montreal, Ottawa, Georgian Bay and Sault Ste. Marie. This rail line began simply as a line from the Laurentians to Montreal. The mastermind of the project was a Roman Catholic priest, François-Xavier-Antoine Labelle. His purposes were to colonize the north shore of the Ottawa River and to develop industry in the area. In particular he wanted the line to be able to ship firewood to Montreal. The city had suffered several bitterly cold winters. He personally was unable to raise the necessary funds so turned to Montreal financier Hugh Allan for assistance.

Allan was a Montreal based shipping magnate and possibly the richest man in Canada at the time. Allan and his fellow investors took the Labelle plan and expanded it. They sought and received funding from the Province of Quebec and the city of Montreal. They had surveys done by engineer Charles Legge of a rail route that would skirt the North Shore of the Ottawa River, enter Ontario at Mattawa and cross Ontario parallel to Lake Nipissing and the French River. The outlet of the French River at the French River Village would give the railway a port on the Great Lakes. The advantages of a railway to ship construction equipment, men and materials to build the canal were considerable. In the section along the Mattawa and French Rivers the railway relied heavily on the previous surveys of Shanly and Clarke. This plan turned out to be a thinly veiled attempt to lure investors into a transcontinental railway scheme.

Allan had his sights set on being the first to build a transcontinental railway across Canada. The Conservative government of Sir John A. Macdonald had in 1871 promised British Columbia such a rail line as a condition of the province joining Confederation. It was to be built in ten years. Macdonald and his Conservative government needed campaign funds for their re-election bid in 1872 and sought out Allan for donations. He did not disappoint, proffering up over $350,000, some of which was provided by American investors. Not surprisingly the Conservatives won the election and Allan was awarded the contract for what was then called the Canadian Pacific Railway.

All was quiet until while Allan and his lawyer, John Abbott, were in London seeking additional financing for their railway, Abbot's private secretary, George Norris, stole letters from Allan's safe. The letters, which proved that Allan was promised the rail contract in exchange for campaign contributions, were sold to the Liberals for $5,000. The Pacific Scandal was born! Macdonald and his government were forced to resign on November 5, 1873 and were replaced by the Liberals under Alexander Mackenzie. Allan's transcontinental railway scheme was shelved. It was not until seven years later in 1881 that another group of investors would successfully launch the transcontinental Canadian Pacific Railway.

8

THE COMPETITION

The United States, and New York State in particular, had developed a system of canals as early as 1832 to ship products from the Great Lakes to the Atlantic. The Erie and Oswego canals opened up a route from Buffalo to New York via the Erie Canal and the Hudson River. This was initially a barge canal that cut transportation costs by ninety-five percent over existing overland routes. In 1918, it was enlarged as the New York State Barge Canal, and combined with the Oswego, Champlain, Cayuga, and Seneca Canals, it provided a 525-mile (845 km.) canal system throughout the state. The economic impact to cities along the route and to New York City itself was enormous. It propelled New York City to predominance amongst east coast Atlantic ports.

Erie Canal System

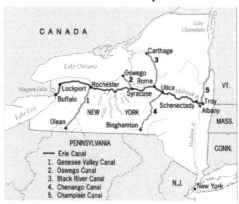

Chicago, center of trade for the American Midwest, had its own plans for economic greatness. In the latter half of the nineteenth century, it also had a water quality problem. Sewage from the city was dumped into the Chicago River and accumulated in Lake Michigan, creating a potential health issue. The city's drinking water was taken from the lake. City officials decided to reverse the flow of the Chicago River and flush the sewage through a canal down the Des Plaines River and into the Mississippi River System. The Chicago Sanitary and Ship Canal was also seen as a means of shipping products through the Mississippi system to the Gulf of Mexico and from there to international destinations. There was a canal in existence, the Illinois and Michigan Canal, built in 1848 for the purpose of shipping bulk items, but it was too small to carry larger freight boats. The Chicago Sanitary and Ship Canal was constructed between 1889 and 1900 with a depth of twenty-four feet and size large enough to accommodate Great Lakes vessels. It was considered an engineering marvel of the time.

This work may have suited the purposes of the city of Chicago but created problems both upstream and downstream. It diverted large amounts of water out of the Great Lakes and caused problems with water levels for shipping. In 1925, the diversion was blamed for Lake Huron having the lowest water levels in sixty-five years. It also drew the ire of Canadian hydro-electricity proponents such as Sir Adam Beck, founder and chair of the Ontario Hydro Commission, who saw the diversion as having a direct impact on the ability of Niagara Falls to produce hydro power. Beck estimated that the diversion of water through Chicago was ten thousand cubic feet per second and cost Niagara Falls 350,000 horsepower in electricity generation per one thousand feet per second of diversion. Downstream states such as Missouri objected to Congress over the treatment of their waterways as Chicago's sewage dump. In addition, Chicago was unique among U.S. cities for not treating its water with chlorine before releasing it into local water bodies. The result was water with high fecal coliform colonies. Signs along the canal indicated that the contents of the water were not suitable for "any human body contact." Litigation from Missouri began in 1907 and continued until 1930. The canal was subsequently turned over to the United States Army Corps of Engineers. The Corps of Engineers reduced the flow of water out of Lake Michigan but kept the canal open for shipping.

The railways that had sprung up in each corner of Canada were legitimate competition for canals. Beginning in the mid-nineteenth century, rail lines were built to connect all parts of the Canadian frontier. Most lines were government-financed through subsidies, interest guarantees, and in the case of the Canadian Pacific Railway, gifts of land. Canadians of all walks of life championed the coming of the railway, especially if it ran through their hometown. The rail companies did successfully compete for government funding with canals, but they could not compete in the price for carrying bulk goods. This prompted a number of schemes, some canals, and some a combination of railway and marine shipping.

A serious rival to the Georgian Bay Ship Canal was the combination of lake shipping and railway transport from ports on Georgian Bay and Lake Huron. These ports were served with rail lines to ports such as at Montreal. This combination had the advantages of a reduced distance of overall travel from the Lakehead to Montreal and the ability of rail lines to operate year round. Ports such as Collingwood, Midland, Goderich, Meaford, Owen Sound and Depot Harbour flourished in the late nineteenth and early twentieth century.

Depot Harbour was built by the Ottawa lumber magnate, J. R. Booth, specifically to take advantage of the growing grain trade from the west. He had built the Ottawa, Arnprior and Parry Sound Railway beginning in 1892 to ship lumber from his holdings in present day Algonquin Park to link with his Canadian Atlantic Railway (CAR). The CAR had its terminus on the St. Lawrence River. It was a natural and potentially profitable extension of his business to build a grain terminal on Parry Island in Georgian Bay and ship grain on his rail lines to the St. Lawrence. The Parry Island site, several kilometres from Parry Sound, was chosen for its fine natural harbour and because the landowners in Parry Sound were asking too much for their port lands. Booth was known as a strong willed individualist and one of Canada's most successful capitalists. He, however, ignored the presence of the Parry Island Reserve and its occupants in building his railway and terminal. Booth repeatedly flouted government regulations and showed disrespect for Indigenous people throughout his career.

Beginning in April of 1898, Depot Harbour was an immediate shipping success. In that year alone one hundred and five vessels carrying over ten million bushels of grain transferred their cargoes to waiting rail cars. Booth focussed his energies on operating an efficient rail line that shipped grain to Montreal in record time. In 1899 it handled 13.6 million bushels of western grain to second place Midland's 6.8

million bushels. Depot Harbour dominated the Georgian Bay shipping scene until well after the sale of the terminal and railway to the Grand Trunk Railway (GTR) in 1904.

Despite the efforts of railways, shipping and grain elevator companies, the Canadian system for handling prairie grain proved inadequate. Canadian ship owners could not keep up with the demand for vessels. As a result the American Great Lakes fleet was engaged. This meant that a percentage of Canadian grain was shipped through American ports such as Buffalo. Canadian ports were overwhelmed by the supply of grain during harvest season. The inadequate elevator capacity and lack of railway rolling stock meant long delays for unloading. As an example on November 2, 1906, the GTR informed shippers that its Georgian Bay ports could not accept any more grain that year. The lack of elevator capacity continued through the 1920's as the capacity of elevators at Port Arthur and Fort William was more than three times larger than that of Great Lakes ports that loaded grain on to rail cars for export. At the time the Welland Canal was woefully inadequate to handle large grain carrying freighters and the Georgian Bay Canal was still but a dream.

An idea that oddly enough came to fruition, though later in the twentieth century, was the Hudson's Bay Railway. Grain was shipped by rail and loaded onto ocean-going vessels at Churchill, Manitoba, for direct transport to Britain. The railway successfully competed for government funding with the Georgian Bay Canal. The drawback of this system was that ships could only ply the waters of Hudson Bay for two to three months of the year. In addition, the timing of the ice-free season on Hudson Bay did not match the most productive grain harvest months in Canada of October and November. Regardless, the government of Robert Borden fully financed this sparsely used line.

The canal system that received the most attention and financing was the St. Lawrence Waterway. It had the twin advantages of a large population along its route and that the cost of construction could theoretically be shared between Canada and the United States. The U.S. was strongly in favour, since its generous neighbour did not charge tolls for its usage, even though the vast majority of canals went through Canadian territory! This route had ample open water for ships to gain speed and save time and expense.

Work began on the Welland Canal in 1829 and was enlarged and improved twice prior to 1890. The St. Lawrence canals were first opened in 1846 and improved in 1890. The St. Lawrence route had the political backing of cities such as Windsor, St. Catharines, Niagara Falls, Welland, Toronto and Kingston which would prove vital to its success in attracting government support.

As early as 1870 the newly minted Dominion of Canada had begun to examine its system of canals by order of a commission. This commission was to study and report on a total of twelve canal projects, including the Georgian Bay route, which it named the Montreal and Lake Huron system of navigation. The commission report stated:

> *If we look at the routes of all other projected canals ... we see that each and all are intended to be subsidiary to the St. Lawrence route. Our duty is to improve that navigation in the first place, because it is the one that has been tried and found to answer all the purposes for which it was intended. It would be unwise to spend millions of dollars of public money in assisting enterprises of minor utility at present, when a comparatively reasonable sum can so improve existing works, like the Welland and St. Lawrence system of canals, as to answer all the requirements of trade for many years to come, and with the certainty of retaining a large income to the public revenues, and giving impulse immediately to the development of the commerce of the whole Dominion.* [25]

CANADIAN GOVERNMENT PUBLIC WORKS PROJECTS 1870

1st The Welland Canal and the enlarement thereof

2nd The St. Lawrence Canals and the enlargement thereof

3rd The deepening of the channels through the rapids of the River St. Lawrence

4th The deepening of the said river at its shallow parts

5th The Rideau Canal and its improvements

6th The reconstruction of a Canal at Sault Ste. Marie between Lake Superior and Huron

7th The reconstruction of a Canal between the St. Lawrence at Caughnawaga and Lake Champlain

8th The improvement of the river Richelieu and Lake Champlain

9th The construction of the Montreal and Lake Huron system of navigation via the Ottawa and French Rivers

10th The construction of the Georgian Bay Canal connecting the Georgian Bay with Lake Ontario

In a similar fashion to the previous commission, a series of questions was asked about each project. The respondents were chosen from Boards of Trade, MP's, town councillors, and businessmen from Ontario, Quebec, and several American cities. The regional variation in answers pinpoints the biases of various corners of the new nation.

To the question of how the Ottawa route would impact the commerce of the Dominion, J.H. Ingersoll of St. Catharines replied that the route was too twisting and tortuous to allow a ship to make decent time and therefore should not be considered. The Honourable Malcolm Cameron from Ottawa claimed that it would open up settlement for 50,000 individuals, develop new timber and water power resources, and due to its considerable shorter length would command a great portion of trade from the western United States.

In comparing freight rates to be anticipated along the Ottawa route versus the Welland Canal, the Ottawa Board of Trade believed there would be a savings of fully one third. The Windsor Board of Trade believed the Ottawa route could not compete.

In terms of which canal, given a similar draught, would be the best investment, James Little of Toronto referred to the advantages of the shorter Georgian Bay route, water power and timber resources and the safety of the route due to sailing on smaller bodies of water. His ideas were strongly supported by businessmen from Quebec. The boards of trade of Windsor, St. Catharines, Hamilton, and Toronto simply stated that the Welland Canal was the superior choice.

In answer to the question of whether tolls on the Ottawa Canal would be sufficient to pay for its construction costs, the Board of Trade for Ottawa believed construction costs on the Welland route to be $40 million versus $25 million for the Ottawa, thus the advantages of cost, time, and freight being in favour of the Ottawa route. The town council of St. Catharines was confident that tolls would not pay for the competitor's construction costs. Not all respondents exhibited such a one-sided and partial view, but by far the majority were supporting the route that passed closest to their city.

The result of this commission was not wholly discouraging to the proponents of the Ottawa route. They did not get their canal but did get monies for improvements to the Lower Ottawa River in the form of the enlargement of the Carillon and Chute-à-Blondeau canals and a lock at Culbert Rapids.

9

THE LOBBYIST

The years that followed the 1870 commission were relatively quiet as railways took precedence over canals. The CPR did privately commission E.P. Bender to report on the Ottawa route in 1879. He reported favourably. In 1891, the Montreal Harbour Commission publicly stated its support of the route and demanded yet another commission to investigate. It was not until 1894, however, that lobbying for the route was ramped up to any great extent. The reason was in the person of a highly energetic and well connected former mayor of Ottawa, McLeod Stewart.

McLeod was the son of William Stewart, a successful lumberman in the Ottawa Valley. William was also a MLA for Ottawa from 1844 to 1847 and during his time in office was a proponent of the Ottawa route. William had large land holdings in Ottawa named Stewarton, south of Gladstone Avenue in what is now East Ottawa. McLeod inherited much of this land, which put him on a sound footing to start his career. He had obtained a BA and MA from the University of Toronto and trained as a lawyer. He ran successfully for mayor of Ottawa in 1887–88 and again in 1889. Although receiving enough votes in the second election, he was convicted of election fraud and banned from future public office.

McLeod Stewart then turned to real estate and built the Molson Bank building in Ottawa and secured the land and plans for the Château Laurier under his company name the Chaudière Hotel Company. He sold the land and plans to the Grand Trunk Railway for their magnificent hotel. He also dabbled in mining by owning and subsequently selling the Anthracite Coal Company to British interests for a profit. He was for a time the president of the Dominion Savings and Loan Company and the Canadian Atlantic

Railway. McLeod was well connected in the Ottawa community as president of the Scottish Presbyterian St. Andrew's Society and a member of the like-minded Caledonia Society. In 1910, he published a book on the history of Ottawa entitled *Ottawa: The First Half Century*.

McLeod Stewart, Mayor of Ottawa, 1887-1888.
Library and Archives Canada / C-002050

Between 1893 and 1917, Stewart was the chief advocate for the canal. He had support in his advocacy from a wide range of individuals, including past governor generals of Canada, businessmen, boards of trade, MP's, civic officials, and British capitalists. In 1893, he delivered an address to the Ottawa Board of Trade entitled "Ottawa an ocean port, and the emporium of the grain coal trade North-West." In it he outlined the advantages of the route, including a new twist that was the advent of Nova Scotia coal as a return cargo for grain carrying ships. He also predicted a savings of two and a half cents per bushel of wheat over the cost of shipping via the Welland route. Stewart was an imperialist with strong ties to Great Britain. He was also imbued with an unrelenting enthusiasm for the new nation of Canada. He occasionally went a bit overboard in his eagerness. As an example, in one of his speeches he predicted an area of fruit farming in Northern Ontario of one million acres, which would surpass the size and productivity of the Niagara Peninsula. He more accurately predicted vast reserves of spruce for a pulp and paper industry, as well as ample water power along the Ottawa River for the development of industry.

Stewart sought and received an Act of Incorporation from the Canadian Parliament in 1894 for the Montreal, Ottawa, and Georgian Bay Canal Company. The company was given a charter to build a navigable channel of not less than nine feet of draught subject to plans being approved by the government. The company could issue bonds of up to $30 million and could sell any surplus hydro power generated. It was stipulated that the company must spend $50,000 within the next two years and finish the canal within eight years. The directors of the company, and there were many, were present and former politicians and businessmen from Ottawa, Port Arthur, Pembroke, Arnprior, Aylmer, Montreal, and other destinations along the route. The practice of applying for charters was a method of getting first refusal on potentially lucrative public works projects. It was a management technique used in particular for ventures requiring large outlays of capital such as bridges, canals, and rail lines.

Stewart wasted no time in hiring the construction firm of S. Pearson and Son of London, England to build the canal. The Pearson firm was world-renowned, having built docks in Southampton and Halifax, the Hudson River Tunnel in New York State, and the Grand Canal in Mexico. Stewart then asked the government to guarantee four percent interest on the company's bonds for twenty years and explained how a fifty cent per ton toll on goods shipped through the canal would not only pay the interest on the bonds but also provide for maintenance, a sinking fund, and dividends to bond holders. He sought approval for his plan from Prime Minister Sir John Thompson, a fellow Conservative whom he believed was an ally. Unfortunately for Stewart, Sir John passed away in December of 1894, the same year as the bill was passed. Stewart continued insistently writing letters to influential businessmen and politicians and open letters published in the press outlining the advantages of the Georgian Bay route. In a letter to the president of the Canadian Pacific Railway, Cornelius Van Horne, he received the reply that "A canal by way of the Ottawa River and Lake Nipissing to Georgian Bay would not, in my opinion, injure the Canadian Pacific in the least."

The following year, 1895, the Canadian and American governments created the Deep Waterway Commission to examine the various shipping routes for products of both nations. The commission understandably ruled in favour of the Welland route, since most of the work would be completed in Canada and most of the resulting traffic would be American.

The St. Lawrence-Welland route was not the preferred route of Marcus Smith, chief engineer of the Canadian Pacific Railway, who had written a report in 1895 on behalf of his employer. Smith's report was more of a review of the work done earlier by Shanly and Clarke. It appears to have been written with the idea of connecting the canal with the Canadian Pacific rail line, as well as using the CPR as a conduit for supplies and workers for the canal. He envisioned the link with the canal at Cantin's Bay, some twenty-five miles (40 kilometres) up the French River. Smith estimated the cost to build the canal was $14,500,000. He updated the previous arguments put forward to include the availability of a new supply of grain coming out of the Canadian Prairie Provinces and minerals that had recently been discovered in Northern Ontario. He also envisioned the huge hydro power resources that could be provided along the canal, not solely for its operation but also for industry. He believed there to be enough power to electrify portions of the CPR.

"The underlying fact is, that it is better for the community, and hence for the railways that serve the community, to turn over to the waterway the transportation of bulky raw material, in which the weight is very large in proportion to the value: and in the transportation of which speed is not essential."[26]

In keeping with the Deep Waterway Commission, he compared the route to the Welland and St. Lawrence route, as well as the Erie Canal and Hudson River. He found the Georgian Bay route to be shorter, safer, faster, cheaper to ship bulk goods, and cheaper to build.

Stewart began in 1896 to lobby newly elected Prime Minister Wilfrid Laurier. He wrote letters to Clifford Sifton, Minister of the Interior and A.G. Blair, the Minister of Railways and Canals, as well as Laurier, describing the vast riches of timber and minerals that could be developed using the Ottawa Canal. In 1896, he organized a large delegation to meet with federal ministers to promote the canal.[27] No ministers showed up to the meeting. He took a voyage to England to seek out his friends, the former governors general of Canada, Lord Lansdowne and Lord Lorne, and received their support for his plans. Once again, he wrote Laurier. Each letter he wrote did not generate a reply other than that it had been received. Laurier was stonewalling Stewart, perhaps because Stewart was a Conservative political supporter, or perhaps because Laurier had other projects in mind, in particular a second transcontinental railway. In true Canadian political fashion, Laurier did appoint a Senate Commission in 1898 to investigate the feasibility of the canal. The commission was headed by Senator Francis Clemow, a Montreal, Ottawa, and Georgian Bay Canal Company member, and senator Charles E. Casgrain from Windsor.

Once again, this committee interviewed people in the know about various aspects of the canal. The first was S.A. Thompson, former secretary of the Duluth Board of Trade. His advice was that the grain and mineral trade was expanding so quickly that there would be ample business for the route, as well as the Oswego and Erie canals in the United States. Major General Sir William Julius Gascoigne, senior army officer of Canada, spoke of the importance of the route as a military highway running entirely through Canada. Marcus Smith, the engineer who had made his own survey of portions of the route, indicated that the issue of water supply at the summit could be worked around. He postulated that a canal of twelve feet depth could be built for $12 million and a fourteen-foot canal for $25 million. Orman Higman, chief dominion electrician, told of the great water power potential of the route, stating that the route could produce more electricity than Niagara Falls. Finally, James Meldrum from S. Pearson and Son said his company would take on the project if the government would guarantee the interest on its bonds. The amount of interest guarantee requested was $340,000 over twenty years.

Prior to the release of the Senate report, two influential members of the House of Commons rose to speak in favour of the project. W.J. Poupore and N.A. Belcourt spoke of the advantages of the plan Stewart had proposed. Both were Quebec politicians, Belcourt a Liberal and Poupore a Conservative. Belcourt also praised Stewart for "his tenacity, splendid courage and persistence."[28]

The senate report was released in the summer of 1898 and strongly supported the plans submitted to build the canal. Terms used were that it was "feasible and practical" and "of great commercial advantage to the trade of Canada." The water power to be produced was seen as a great boon to industry along the route. Once again, it appeared to be time to build.

Americans were alerted to the dangers of losing trade to Canada in an article in the *North American Review* written by J.A. Latcha. "When the British Northwest can raise and ship by canal one hundred million bushels of wheat, British capital will build the Georgian Bay Ship Canal, and every ton of traffic from the Lake Superior regions to the ocean will traverse British territory, leaving Detroit, Toledo, Cleveland and Buffalo hundreds of miles from the direct route to the ocean."[29]

The Senate report was followed up by a business synopsis written by Stewart and a letter of support written by Poupore. Both were sent to the prime minister. Poupore asked the government to guarantee a stock issue with interest of two percent for twenty years on $17 million worth of bonds. Stewart also approached Laurier with a plan to build the canal for an agreed sum if the government would pony up half the cost in cash. Laurier's response was lukewarm at best. He wanted to know if the canal could be built for $17 million and with a reasonable certainty of commercial success. He indicated that further investigation was required. The prime minister who was known for his "sunny ways" approach to governing was throwing a wet blanket over Stewart's enthusiasm and effort.

In answer to the question of why Stewart failed in his attempts to get the canal constructed, there are several potential reasons. Stewart was a Conservative supporter with strong ties to the monarchy. He named among his friends several past governor generals and had done business with British capitalists. Laurier, in contrast, was attempting to build a united Canada by equally supporting both French and English and wished to create a Canadian identity separate from that of Britain.

Stewart was also a somewhat unstable individual who, despite his many talents, was to spend a number of years in the asylum at Verdun. He claimed to have the neurological disorder Grave's disease, but Mackenzie King had other thoughts. King, the lifelong teetotaller, wrote in his diary, "It was prosperity that was harder on him than adversity. He is in the asylum today, his brains wasted by liquor."[30] Stewart may also have been over zealous in his approach to the extent that Laurier tired of his letter-writing and lobbying.

Stewart initially gave up his quest in 1899. He had successfully negotiated renewals of the charter for the Ottawa, Montreal, and Georgian Bay Canal Company in 1896 and 1898. In 1899, he sold the charter to George Grote Blackwell of England, who immediately flipped it to the New Dominion Syndicate and its manager, James Malcolm. He retained shares in the syndicate until a squabble over expenses led him to sell his shares to Johnson Edgerly. At that time, Stewart disappeared from public life, spending almost two years in the Verdun asylum.

The man and money behind the New Dominion Syndicate was British politician Sir Robert Perks. Perks was the son of a Methodist minister who first qualified in Britain as a lawyer specializing in parliamentary and railway matters. He abandoned law for a career in engineering and was associated with building dockyards and the Manchester Ship Canal in Britain. He entered politics in 1892 as a liberal in the riding of Louth and remained in the British House of Commons for eighteen years. Perks took a strong interest in the idea of building the Ottawa Canal and on several occasions offered to build it for Canada. He was so expert in his knowledge of the plan that a paper he presented to the Royal Society of the Arts Colonial Section won him a silver medal. He was to remain in the Ottawa Canal picture from 1899 until its demise in 1927.

The New Dominion Syndicate took heed of Laurier's advice for further investigation and released funds for another survey of the summit region of the canal. This was completed by George C. Wisner, a consulting engineer from Detroit, who gave the plan a thumbs-up. In 1903, once again, a three-person panel was appointed to review the economic feasibility of the route. The three-man panel of Bertram, Reford, and Fry, dubbed the Transportation Commission, reported in 1905 that the building of the canal was an urgent matter. Again, the report was buried. A frustrated MP, George Monk, asked aloud in the House of Commons, "What new method can the government adopt in order to shelve the matter once more."[31]

The answer, albeit beginning a year previous, was another survey. In 1904, Blair's replacement as minister of railways and canals, Henry Robert Emmerson, along with Minister of Public Works C.S. Hyman, commissioned a full and detailed survey.

Emmerson was a former premier of New Brunswick whom Laurier had recruited to national politics. Emmerson fully indulged himself in the nightlife of Canadian cities, both in drink and female companionship. In 1906, he pledged to the prime minister that if he had another drink, he would resign from cabinet. Unfortunately, in 1907 he was thrown out of a Montreal hotel in the company of two women of ill repute and was forced to resign.

10

THE 1904 TO 1908 PUBLIC WORKS SURVEY

"*I want you to understand that I am wedded to the Montreal, Ottawa and Georgian* Bay Canal; that is a work that must be undertaken in the near future, but Canada is big enough and wealthy enough for both (it and the Welland Canal) ... When the National Transcontinental (Railway) is completed — and that will be before many years, for by 1913 we expect to have the National Transcontinental finished from the Atlantic to the Pacific — then Canada will be in a position to have both these great canal systems complete."[32]

—William Pugsley, Minister of Public Works in the Laurier cabinet

The tidy sum of $250,000 was put toward the task of a complete survey of the route, which eventually cost $694,000 and took a full four years. It was considered one of the finest engineering reports of its time, and even today the 691 pages are surprisingly readable.

There were four engineers heading up the survey; Eugene D. Lafleur was chief engineer and A. St. Laurent engineer-in-charge. St. Laurent was supported by two regional or district engineers, C.J. Chapleau and C.R. Coutlee. The route was divided into three main sections called the Montreal, Ottawa, and Nipissing. In each section were a number of smaller divisions called reaches, roughly twenty-four in total. These typically covered thirty five to forty miles of the route. For each of three reaches, there was a sectional engineer, two assistant engineers, two rodmen, two chainmen, one foreman and seven to nine labourers. The crews worked year-round in dangerous conditions, only losing two men to drowning throughout the four-year span.

The survey looked at absolutely every angle of the route, including the history of the region, past surveys, current and future economic development, and the history of the Canal Company itself. The effects of the proposed work on lumbering along the Ottawa River and flooding of properties created by dams were considered.

In the physical sphere, information was collected on hydraulic characteristics of the rivers, including discharge measurements; the climate of each main section including rainfall, snowfall, and evaporation data; watershed information, including size and depth of water bodies and currents; plus an analysis of rock and sediment borings. The region was also surveyed with precise levelling.

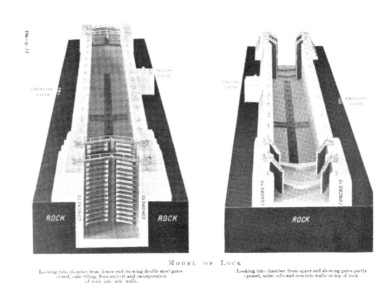

Models of locks

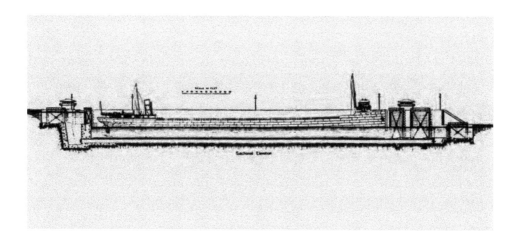

Freighter position in a lock

In the engineering realm, research focused on the locations and dimensions of dams and canals, with detailed drawings of forty-five dams, twenty-seven locks, and thirteen bridges to be built along with the sixteen existing bridges on the route. The locations, horsepower, and uses of hydro dams were noted. The details of the locks and their operations were carefully sketched and described, and comparisons were made with locks around the world. As a part of the venture, two of the survey engineers were sent to Panama to view the engineering work of the Panama Canal.

The initial survey was designed to use hydro power to power locks and to provide lighting along the route for nighttime travel. There were an estimated ten power dams along the route and the potential to produce one million horsepower of electrical power.

The survey produced maps of each section of the route, three large maps of the route, one topographic map, two railway maps, and three drainage basin maps. These were of excellent quality, as can be seen in the example below.

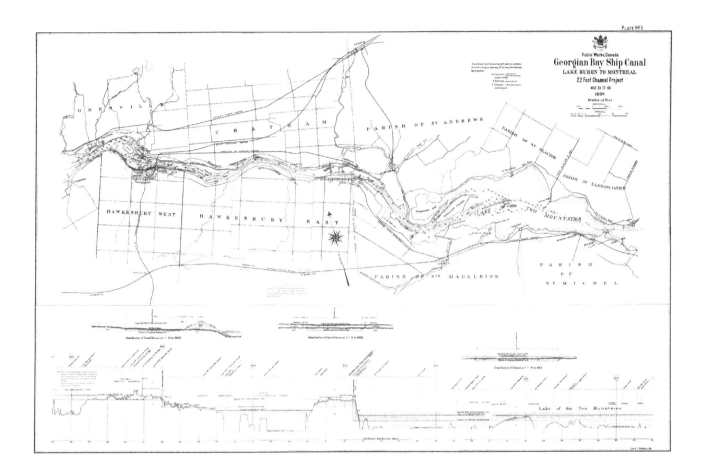

Survey map of Hawkesbury

In practical terms, Great Lakes ship captains were interviewed about the space needed to manoeuvre lake boats and safety conditions on the route. The characteristics of lake boats were examined, as were previous mishaps in the St. Mary's River and St. Clair River. As in previous surveys, the distances between major cities along the route were calculated and compared with other shipping schemes.

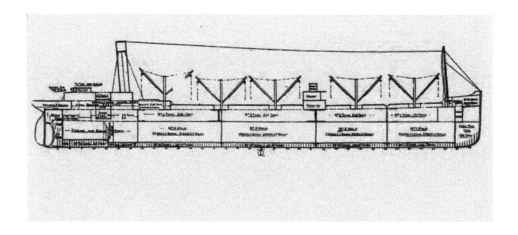

Plan of freighter for the Georgian Bay Canal

Trade statistics were sought and published for goods being shipped along the Great Lakes waterways. The trends showed a healthy growth in the Canadian grain trade in the first few years of the first decade of the twentieth century. Eighty percent of this grain was destined for Britain, with Germany the second largest importer. They also showed that thirty-two percent of Canadian grain was carried in American ships and exited through Buffalo and the Erie Canal. In 1904, the first year of the survey, there were an estimated 280 Canadian ships on the Great Lakes, as opposed to 519 American ships. Imports to Canada were coming in via American ships.

The many mineral discoveries within close proximity to the route were highlighted including silver at Cobalt, copper at Temagami, nickel and copper at Sudbury, and iron ore at Atikokan. Of great national interest was Nova Scotia coal from Cape Breton, which would furnish a substantial inbound cargo for Canadian ships. This, coupled with iron ore from Lake Superior and Quebec, could greatly enhance Canada's steel industry. At that time, coal from the Ohio Valley was being shipped into Canada in American bottoms.

The costing of the route was completed with each section being costed separately then totalled. Two alternatives to the approach to Montreal, named the Montreal Reach and the Lake St. Louis Reach, were considered and costed, with the latter costing approximately $6 million less. The annual maintenance costs of the route including projected staffing of one hundred were also calculated. This work of engineering came complete with several native legends from along the route.[33]

Space does not allow for a full account of the survey findings, all 691 pages of them. Here are some highlights beginning at the Georgian Bay end of the route. The French River empties into Georgian Bay in three outlets. Only one of the three, the middle channel, was determined to be fit for navigation. The harbour was planned to be at the French River Village and had an average depth of thirty feet.

The village was a centre of the French River lumber industry from the 1870s until the 1920s. Logs from upstream as far as Lake Nipissing were floated down the French River in the spring. Booms of logs were built and towed by steam tugs as far as Michigan, where there was a huge demand for Ontario pine. There was also a sawmill at French River Village where pine timber was sawed by the Ontario Lumber Company and shipped by boat to various Great Lakes ports.

The approach to the harbour at French River Village was seen to need some excavation of large rocks and would possibly need further survey work. The harbour itself was well protected from onshore winds by the Bustard Islands and was free from fog for all but four days per year. The French River climbs sixty-two feet from Georgian Bay over a distance of sixty-three miles. In general, it is a wide body of water enclosed by granite cliffs on either side.

The French River delta

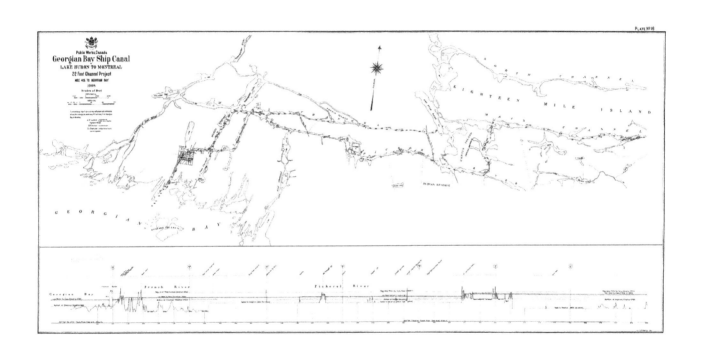

Map of French River outlets

Georgian Bay shoreline

Once entering the French River, ships would encounter the Dalles Rapids, which would be managed with a lock, the first on the route. Farther along the river the main channel contains numerous bends. Critics of the route termed this section "serpentine." The engineers thus chose to divert traffic at Cantin's Island via the straighter Pickerel River and then rejoin the French at Horseshoe Falls. Horseshoe Falls was aptly named and would require some cutting to allow lake boats to make the bend. Engineers were concerned about the difficulty in cutting through Precambrian rock, so they built these costs into their estimates.

THE GEORGIAN BAY SHIP CANAL

Three views of the French River from the 1908 survey

The next obstacle was the Récollet Falls, then Five Mile Rapids, a series of four different rapids, which would require additional cutting to create a canal. Above Five Mile Rapids, it would be clear sailing to Chaudière Falls, where a dam and lock would be necessary. These falls are at the end of the west arm of Lake Nipissing. Much of the outflow of Lake Nipissing travels over these falls, so by damming the river at Chaudière, the plan called for increasing the level of the shallow Lake Nipissing by eight feet above the low-water mark. This would flood some lakeside farms, but again, the cost of purchasing this land was accounted for. Engineers were again aware of the need for heavy rock-cutting at Chaudière. Since the French River had an annual flood range of eight to ten feet, the water control measures would alleviate some of this difference. Once through the lock at Chaudière, ships had sixty-five kilometres of clear sailing on Lake Nipissing.

Rapids at Chaudière

Four miles east of Lake Nipissing is the divide between waters flowing west into Georgian Bay and east into the Ottawa and St. Lawrence Rivers. A canal would be cut through this divide to join Lake Nipissing with Trout Lake. Trout Lake, Turtle Lake, and Talon Lake form the headwaters of the Mattawa River. They are all navigable lakes, but there is a drop in elevation of 180 feet (55 metres) between Trout Lake and the Mattawa River requiring the building of a series of locks.

The Mattawa was considered navigable as a channel with high rock walls for its full length until it joins the Ottawa. It was not without its rapids over its sixty-five-kilometre length, there being eleven canoe portages in total. The engineers believed they could funnel enough water through the Mattawa basin to drown these shallow rapids.

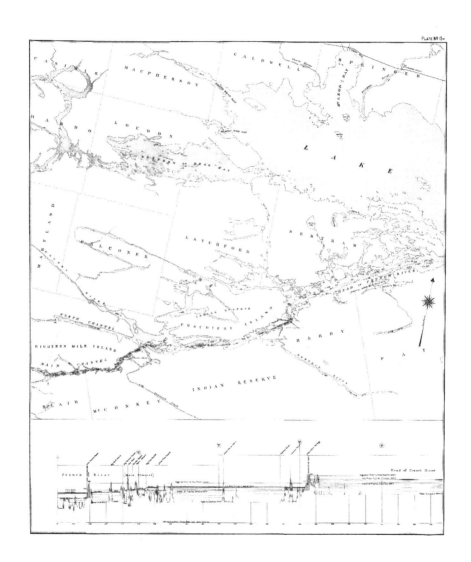

Lake Nipissing map from 1908 survey

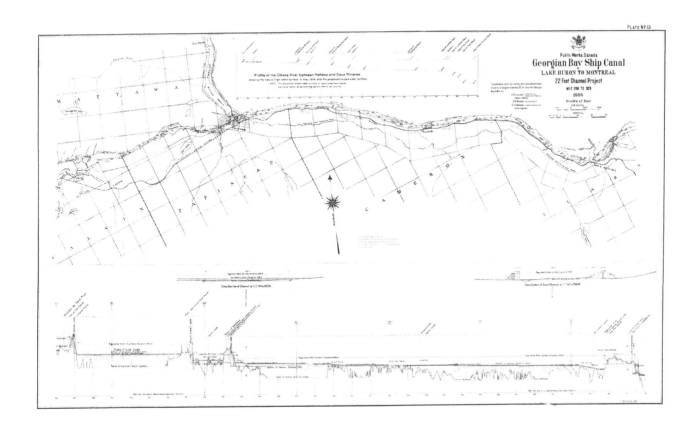

Map of Ottawa and Mattawa Rivers

The Ottawa River was the gem of the route. It covered some 360 miles (579 km.) of the route's 440 miles (708 km.). It was encased in a wide valley, the former glacial spillway, and was wide enough and deep enough for lake freighters along most of its length. It was essentially a series of deep, wide basins separated by falls and heavy rapids. The two additional characteristics of the river were its volume and its speed. It rose in the divide between the Hudson's Bay watershed and the St. Lawrence watershed and travelled some 750 miles (1207 km.) to the St. Lawrence. Over this span, it dropped 1,045 feet (319 metres) in elevation. The river's discharge varied greatly from spring freshet through fall and was occasionally prone to flooding. In 1876, the river experienced its largest flood, destroying mills, logging operations, and farms along its shores. This flood recorded a height of twenty-five times the low-water record and was estimated at Ottawa to be discharging 300,000 feet per second of water. The engineers were quick to point out that this flooding could be controlled by damming the large lakes in the northern Ottawa Valley that fed the river.

Ottawa River near Grenville

At the time of writing of the report, there was navigation for passengers and freight between Ottawa and Montreal. John Molson of Molson Brewery fame had put the first steamship on the Ottawa River, and there was regular service established between the two cities. There had been locks built at Grenville, Carillon, and St. Anne's to allow for steamboat traffic. In addition, there were sections of the river that supported the transportation of logs. These included Deschênes Lake, Lac des Chats, Colougne Lake, the lower Alumette Lake and the stretch of the river from Pembroke to Rapides-des-Joachims. There were numerous falls and rapids to be overcome north of the city of Ottawa, but with those came with the opportunity to produce vast amounts of hydroelectricity.

The terminus of the route was at the Port of Montreal. In the report, the chief engineer of the Montreal Harbour Commissioners is quoted as saying, "Montreal has behind her a canal and river system fourteen feet deep, tapping the trade of almost a whole continent. Equip in a proper manner her ocean and lake terminals and no force can divert from the cheapest and shortest trade route the business she ought to command."[34] Between 1895 and 1905, shipping at the Port of Montreal had doubled. In 1898, the Laurier government pledged one million dollars to upgrade port facilities. These included concrete quays, steel storage sheds, docks, and grain elevators. Montreal was staking its claim as the premiere port of Canada.

The conclusions of the four-year survey were presented to Minister of Public Works William Pugsley and the House of Commons in July of 1908. The final report was tabled the following January. The report concluded that a twenty-two-foot draft waterway could be completed in ten years at a cost of $100 million. The annual maintenance costs were deemed to be $900,000. In total, the route would be 440 miles (708 km.) in length with twenty-seven locks, eighteen main dams, twenty-eight miles of canals, and 66 miles (106 km.) of channel dredging. The possibility existed to deepen the channels to twenty-five feet as the size of freighters increased. There would be 116 curves along the route. A modern lake freighter would take seventy hours at an average speed of twelve miles per hour to navigate the route. The route would be open to ship traffic 210 days per year.

Port of Montreal, 1899

Great Lakes grain carrier

One hundred million horsepower of electricity could be produced along the route. The report urged the federal and provincial governments to sort out ownership of the lands on either side of the route, the islands in the rivers, and most importantly who owned the hydro power before the canal was built.

While the engineers and their crews were busy with the survey, more lobbying and investigation was taking place. In the first year of the survey, 1904, Laurier and William Mulock received a deputation of boards of trade and municipal councils from Ottawa, Aylmer, Renfrew, Arnprior, Pembroke, North Bay, Mattawa, Sturgeon Falls, Warren, and Cape Bay, urging action on building the canal. They were joined by MPs from Nipissing, Pontiac, Smith Falls, Renfrew, Lachute, and Pembroke.

Another commission was struck to report on the feasibility of the canal in 1905. It was organized by Secretary of State Charles Murphy and included a ship owner, a banker, an engineer, and a manufacturer. This committee strongly urged the immediate start to the canal. The prime minister's reply was that the government would attend to this matter as soon as they had the money.

Macleod Stewart returned from his furlough at the asylum in Verdun to embark on a speaking tour of Boston, Winnipeg, and Minneapolis. The topic was not surprisingly the Georgian Bay Ship Canal. In 1906, he began a series of four open letters to Laurier urging action on the canal. Again, there was no reply other than that the letters had been received.

The charter for the Canal Company, which included the right to build, continued to get passed by the House of Commons or the Senate with slight modifications every two years from 1896 to 1908. In 1908, several rights of the initial charter were removed. These included the right of expropriation along the canal and the right to fix power rates generated by the canal. Also established was the need for consent from the provinces of Ontario and Quebec to sell power produced by the canal.

Laurier in the election year 1908 visited North Bay, and ever the politician, made this statement: "Let me say that this canal is … very much in my heart; it is not a monopoly for the people of Nipissing or Northern Ontario, it interests the people of Montreal living along the St. Lawrence, and the men who are today growing wheat in the far West … The wheat of the West needs the shortest and cheapest route to the sea, and that route is the Georgian Bay Canal… We are not yet ready to build the Georgian Bay Canal because we have a falling revenue and heavy expenditures. But if Providence spares me and my colleagues in power, it will be our duty to take up the Georgian Bay Canal as soon as revenue permits."[35]

11

LAURIER AND THE RAILWAY MESS

LAURIER SO BELIEVED IN THE FUTURE OF CANADA THAT HE FAMOUSLY PRONOUNCED THAT the twentieth century would belong to Canada. "The nineteenth century was the century of the United States. I think we can claim that it is Canada that shall fill the twentieth century."[36] Upon taking office in 1896, he was rewarded with an economic boom in Canada and worldwide that would last until

1913. This was preceded by a worldwide depression lasting from 1873–96 and was coupled with many Europeans wishing to leave the continent to escape poverty. He used this prosperity as an opportunity to pursue his dream of a second transcontinental railway. Railways had sprung up across Canada since prior to Confederation with reckless abandon. In the Province of Canada, for example, two pieces of legislation, the *Guarantee Act* and the *Municipal Loan Act*, had been created to promote rail construction. The response was a mania of railway construction that left financial problems for railways and municipalities in its wake. Sir Wilfrid Laurier was, like his predecessor Sir John A. Macdonald, set on building an intercontinental railway in Canada to compete with the monopolistic, privately owned Canadian Pacific Railway. His railway, as opposed to the CPR, would be an all red-route avoiding crossing into the U.S.

The railway, the National Transcontinental, was to become much more expensive than planned and would eventually contribute to his loss in the 1911 election. He tried several angles to get it built. He first tried to combine the lines of the Grand Trunk Railway in Eastern Canada with lines in the west being built by former CPR builders William Mackenzie and Donald Mann. The two railway entrepreneurs said no to Laurier's offer and proceeded, with bond guarantees from the federal government and a grant of six hundred acres of land per mile of railway constructed, to build Canada's second transcontinental line, the Canadian Northern Railway. Laurier then turned to the staid and moribund British-based Grand Trunk Railway, which had rail lines primarily in Eastern Canada. At that time, it was undergoing a revitalization by hiring American railroader Charles Melville Hays as general manager. Hays was to introduce more modern "American" methods to the railway. Laurier decided that the government should finance a line eastward from Quebec City to Moncton and westward across Northern Quebec and Ontario to Winnipeg. This was to be named the National Transcontinental. The Grand Trunk Pacific (a wholly owned subsidiary of the Grand Trunk) would then build a line from Winnipeg to Prince Rupert on the west coast. When all lines were completed, the railway would be leased to the Grand Trunk Pacific at very favourable terms.

The pursuit of the Grand Trunk Pacific venture brought about the resignation of Laurier's Minister of Railways and Canals, Andrew George Blair, former premier of New Brunswick. He spoke about the reasons for his resignation in the House of Commons for five hours. He was in opposition to the government building and owning the lean section of the railway system in Canada while financially supporting the fat section. "… I am not in favour of impetuously rushing into the construction of a transcontinental line from Quebec, through an unknown country to Winnipeg and the west until we know something about it … The project is one of very great magnitude, and should be dealt with only after the maturest deliberation."[37]

Laurier the skilled orator and ever the optimist responded to the criticism by stating, "… this is the time for action. The flood of tide is upon us that leads on to fortune; if we let it pass, the voyage of our national life, bright as it is today will be bound in shallows … we cannot wait, because at this moment there is a transformation going on in the conditions of our national life which would be a folly to ignore and a crime to overlook."[38] Blair accused Laurier of working behind his back on these railway plans. In order to avoid further criticism, Laurier appointed Blair to the head of the Railway Commissioners, which took him out of the House of Commons. Blair resigned from this position in order to run against Laurier in the 1904 election. After discussions with Laurier, he withdrew his bid for election.

Laurier was hit with another setback in his effort to build the National Transcontinental. As a representative of the riding of Quebec East, he spearheaded attempts to build a bridge across the St. Lawrence River from Quebec to Lévis on the south shore. This bridge was a part of the National Transcontinental that went through Quebec to Moncton. It was to be built by a separate company, the Quebec Bridge and Railway Company. Laurier initiated a guarantee on bonds in the value of $6 million to get the bridge built. It was to be the longest cantilever-style bridge in the world.

It became apparent during construction that the design of the bridge was faulty, as the actual weight of the bridge was far in excess of its carrying capacity. The site engineer, Norman McLure, started to notice in the summer of 1907 that structural members already in place were bending. A workman had reported that a bolt he had inserted into a beam had immediately snapped. McLure informed the consulting engineer for the QBRC, Theodore Cooper, who indicated that the issue was minor. Cooper said that the beams must already have been bent before being added to the bridge. McLure persisted in his complaints and wrote a letter to Cooper and then went to visit him in New York on August 29. Cooper took the visit seriously and cabled the Phoenix Construction Company not to add any more load to the bridge. This message was not relayed to the Quebec site in time, and on the afternoon of August 29, 1907, very close to quitting time, the bridge collapsed, killing seventy five workers. The time it took the structure to completely collapse into the St. Lawrence River into a pile of twisted metal was fifteen seconds. Of the victims, thirty-three were steel workers from the Kahnawake reserve near Montreal.

Quebec bridge collapse, 1907

A Royal Commission of Inquiry was hatched and found the fault for the disaster rested with Cooper and Peter L. Szlapka, chief engineer of the Phoenix Construction Company. It also found the Quebec Bridge and Railway Company to be negligent in not hiring an experienced engineer and in the supervision of the work being done. A second bridge was begun shortly after the report of the royal commission. Three experienced engineers were put in place and a modified design selected. The new bridge fared only slightly better. The central span of the bridge collapsed while being hoisted into place on September 11, 1916, killing thirteen workers. This time the fault was found to be with the hoisting mechanism. The bridge was finally completed in 1917.

The flurry to build two more transcontinental lines brought an unprecedented boom in railway building in Canada, a nation that was already well served by rail lines.

Table 1 – The Canadian Railway System in 1904

- The **Grand Trunk**, the oldest railway, headquartered in London, England, operating in Eastern Canada, based in Montreal, with no extensions West but desirous to go there and with 3,750 miles of track. It had a well-established feeder system in Ontario and Quebec.

- The **Intercolonial**, provided for by an article in the British North America Act, based in Moncton, which united the Maritimes with central Canada by linking Nova Scotia and New Brunswick with Quebec. It operated with 1,300 miles of track plus another 300 miles through subsidiaries.

- The **CPR**, a well-run, Montreal based railway, with 9,000 miles of track from St. John, N.B. to Vancouver, (nearly half the trackage in Canada), operating in both East and West, with a national trunk line supported by regional feeders. Its operations were profitable.

- The **Mackenzie & Mann system**, (to become the Canadian Northern System) headquartered in Toronto with strong ties to the Bank of Commerce. Its nearly 3,000 miles of track were principally on the Prairies between Winnipeg and Edmonton, a part of the country experiencing dramatic growth at the time, although expansion into Eastern Canada was on its wish list. Its operations were profitable.

- There were, in addition, 57 other much smaller companies, ranging in size from the 382.19-mile Michigan Central in Ontario to the tiny two-mile Klondike Mines Railway in the north.

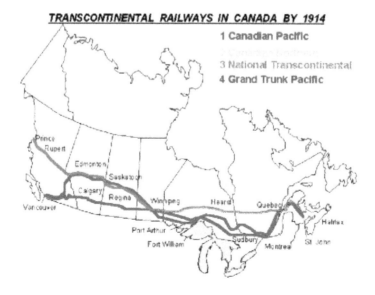

Laurier's government financed the National Transcontinental Railway and supported the Canadian Northern Railway as one of three national transcontinental lines. Clifford Sifton, then Minister of the Interior, came up with a plan to assist the Canadian Northern. In it, the federal government would guarantee the interest on bond issues sold by the company to raise monies for construction. McLeod Stewart had put several similar plans together to finance the Georgian Bay Canal but could not get the necessary government support.

The excess rail capacity in Canada soon became apparent, and the Canadian Northern was nationalized in 1918. It was combined with the National Transcontinental to form Canadian National Railways in 1919. It was left to the Borden government and then Finance Minister Sir William Thomas White to take over these two transcontinental railways and create the CNR, a corporation managed by trustees appointed by the federal government. The Borden government chose to assume all existing interest charges on railway securities issued by the Canadian Northern and Transcontinental lines. The bill for the Canadian Northern alone was some $218,215,409.

When Stewart and others came calling on Laurier or wrote him letters asking for financial backing for their canal plans, he ignored their requests in favour of financing two transcontinental rail lines in a nation of only seven million inhabitants. Laurier had many fine traits as a statesman and nation builder, but his excessive focus on railway-building could be viewed as his Achilles heel.

In the early 1900s, Laurier's Liberals faced political pressure to restart the Trent Valley Canal that had been twice previously abandoned. In 1904, the canal reached Peterborough and Lake Simcoe. The outlet to Georgian Bay through the Severn River was completed in 1920, but with a maximum boat draught of six feet, it was woefully inadequate for commercial purposes.

In 1904, Laurier did commit the government to support the Georgian Bay Canal project. In 1908, he said the work would proceed when the resources of the Canal Company permitted. That was all Robert Perks needed to hear. He travelled to Ottawa in May of 1909 to propose a plan to build the canal with financing from British capitalists. All he needed from the government was a three and a half percent guarantee on the bonds sold to raise working capital. Laurier said no to the guarantee but yes to the work being started with the government's acceptance of construction plans. Finally, Laurier was saved in 1911 by the sound of the election bell from spending any money on the project. He did include it as part of his election platform, as did Robert Borden.

The 1911 election was a heated one by Canadian standards, with reciprocity with the United States being the main issue. Ironically, it was the Liberal platform that proved their undoing. Borden and the Conservatives were able to capitalize on pro-British sentiment, which focussed on keeping Canada's trade ties with the British Empire intact. The fear of reciprocity with the United States, leading to commercial and eventual political takeover, was real and aided the Conservative cause. Borden was backed by a powerful group of Ontario industrialists and business leaders named the Toronto Eighteen including the likes of John C. Eaton, Sir Edmund Walker, president of the Canadian Bank of Commerce, and W.T. White of the National Trust Company. His chances were given a huge boost by disaffected Liberals such as the former Minister of the Interior, Clifford Sifton. Sifton spoke around the country, often on the same stage as Conservatives, on the ills of reciprocity.

Borden and the Conservatives won a majority in the House of Commons in 1911 and served up another target for enthusiastic supporters of the canal. Fortunes seemed to have shifted in favour of the canal with F.D. Monk, a supporter, being appointed Minister of Public Works and Frank Cochrane, the MP from Nipissing, being appointed Minister of Railways and Canals. One of the first to communicate with the new prime minister was the lifelong Conservative, Macleod Stewart. Unfortunately Stewart was destined for another visit to the asylum and was not able to formally meet with Borden. In later years, Stewart would seek the source of rumours that the canal project was largely ignored by Laurier because of Stewart's involvement. He confronted Laurier with this assertion, to which Laurier smiled and said, "The Georgian Bay Canal without you would be like the play of Hamlet with Hamlet left out."[39] *Hamlet* was of course a tragedy.

The fight for the canal was taken up by a group of MPs in constituencies along the route. Two large delegations visited Borden to advocate support for both the Georgian Bay Canal and the deepening of the Welland Canal. Borden's answer was that swift action was not possible and that further study was required. He did, as his predecessor was wont to do, create another commission to study the project. It began in 1914 and was a three-person panel headed by the ex-mayor of Winnipeg, W. Sanford Evans. It was tasked with determining whether an expenditure of one hundred million dollars on the canal was economically feasible. The advent of the First World War delayed the commission's work, and their report was not fully completed until 1917.

Concurrently, a fight for Borden's support was being waged in the private sector amongst boards of trade and municipal councils. The fight was clearly a bi-partisan battle between supporters of the Georgian Bay Canal and those of the Welland Canal.

12

THE GEORGIAN BAY CANAL VERSUS THE WELLAND CANAL

THE COMPETITION BETWEEN CANALS HEATED UP WITH THE FORMATION OF THE "CANADIAN Federation of Boards of Trade and Municipalities" with members from the Maritimes, Prairies, Ottawa Valley and northern Ontario lobbying for the Georgian Bay Canal. Two months later supporters of the Welland Canal formed the Great Lakes and St. Lawrence Navigation Association. A pamphlet produced by The Port Arthur Board of Trade entitled *Read! Investigate! Judge! And Act! The Welland or Georgian Bay Canal Which* authored by Port Arthur's Joseph Redden, was one of many produced to lobby the public and their political representatives.

Redden writes:

"We have also brought before our eyes almost daily, with a greater force than the best of government statistical blue books can give, the great leak in the main artery of Canadian transportation. We refer to the large number of American vessels that leave these upper lake ports of Canada every year, loaded with hundreds of thousands of tons of Canadian freight consigned to European countries, routed via American lake ports, for storage and transhipment by American railways to American seaports. We also see here the importation of hundreds of thousands of tons of American soft coal brought in by the same vessels, and all because the present canal route of Canada from the sea and her eastern coal area is totally inadequate for the passage of large vessels, if the passage of large vessels were possible it would mean a cheaper transportation of freight than the American route could make possible, and it would retain to Canada the routing through her own channels, to and from the sea of the millions of tons of freight which now are routed through American channels."[40]

He continues his convincing arguments:

"There is one fact that must be borne in mind when judging of the merits of either the Welland or Georgian Bay Canal, and that is that the benefits of the canal improved or built will largely go to the country on the shores of which the best storage ports are situated. By the deep Welland Canal the United States have these ports. Some may say Canada has, at the lower end of Lake Ontario, in Prescott and Kingston, as good storage ports as the United States. Not quite. And Canada cannot compete with the United States, for she has not the west bound traffic by this route, which might enable her to build huge vessels and constantly keep them in commission, carrying as cheaply as the United States vessels. The most profitable vessel for Canada, with a deepened Welland Canal, will still be the 2,200 ton canal draught type of vessel suitable for return cargoes of package freight from Montreal and Intermediate ports, practically the only available west bound freight. The Georgian Bay Canal will change these conditions, which exist, and will exist, on the Welland Canal route, to the detriment of Canadian interests, and if, as the Hon. R.L. Borden has said, "our object is to keep Canadian trade in Canadian channels and to continue, as much as we can, the policy of making the trade run east and west," it is the Georgian Bay Canal which should be built, as it alone will enable such desires to become realities."[41]

Redden goes on to document the facts surrounding the size of American vessels versus Canadian. The Americans had enlarged their vessels to take advantage of shipping newly discovered iron ore in Minnesota. They also had a return cargo for these vessels in coal from the Ohio Valley. The newer American vessels could carry up to 12,000 tons, whereas Canadian boats were in the two- to three-thousand-ton range. Advantage U.S.A.! Thus American vessels could and did carry Canadian freight cheaper than could Canadian vessels. This situation would be rectified with the building of the Georgian Bay Canal to a depth that would carry larger vessels. The outbound cargo would be grains from the prairies, the incoming coal from Cape Breton for future Canadian steel mills.

Redden goes on to reproduce in the booklet reactions of the American press to building the Georgian Bay canal.

New York Herald — "No effective competition with this route appears in any way possible. When in operation the Buffalo route will be hopelessly outclassed, and the St. Lawrence will then solve and control the transportation conditions of the continent."[42]

The *Chicago American*—"The proposed Georgian Bay Canal, if placed in operation, will deprive the United States of millions of tons of freight annually, and deeply affect our markets." [43]

The *New York Sun* — "The actual transportation distance from the Soo to New York, by way of Lake Erie and the Erie canal, is about twice as great as that from the Soo to Montreal via the projected Georgian Bay route. It is estimated that the cost of transportation of wheat to tidewater would be reduced by at least two cents per bushel. Between this route and its twenty-one feet of navigable depth, and the $101,000,000 gutter across New York State, the odds as a business enterprise are emphatically in favor of the Georgian Bay canal."[44]

Redden concludes:

"With the Georgian Bay Canal in existence Canada will hold sway as the great factor in the transportation of western freight, both United States and Canadian, through Canadian channels and terminals to the sea, for nearly eight months of each year, and almost equally so for the remaining four, for Canadian freight. The boundary waters of Canada will, and must remain forever subject to the joint control of our neighbors and ourselves. But nature has given us, In the Ottawa and French rivers, the means of providing ourselves with a great, independent, national deep waterway, and most Canadians strongly believe its immediate development to be the only truly national waterways policy for this country."[45]

The counter arguments were many, as outlined in a pamphlet produced by the Great Waterways Union of Canada. This organization with a lofty title was composed of mayors, municipal officials, and businessmen from London, Owen Sound, Galt, Preston, Guelph, Waterloo, Berlin, and St. Catharines. Their treatise was authored by D.B. Detweiler of Berlin and entitled *The Inland Waterways of Canada Ocean Navigation via St. Lawrence and Welland Route Georgian Bay Route Impracticable.*

This lobby group declared their pamphlet to be a protest against the Georgian Bay Canal as a waste of government time and money. They cited the 117 curves along the route making it dangerous and increasing the insurance rates on shipping. The contention was that completing the Georgian Bay Canal would take ten years, and Canada did not have ten years to wait. The State of New York had pledged $100 million for the Erie Barge Canal, and it was to be completed in approximately three years. The argument put forward was if the Welland Canal was begun immediately, it would take about the same time to complete as the Erie Canal. The group urged immediate action to create a St. Lawrence waterway that could accommodate ocean-going vessels similar in size to those being used by U.S. companies.

The group pointed to canal failures such as the Trent Valley Canal and the Newmarket ditch (The Toronto to Georgian Bay Canal) to highlight the importance of not rushing into two projects before completing one. These supporters saw the vast hydro resources available on the southerly route and pointed to the fact that they had a perceived ally in Adam Beck, chairman of the Ontario Hydro Commission and a London resident. In a contentious argument that would reappear years later, they thought that hydro power from the St. Lawrence could be used to pay for canal work on the St. Lawrence route.

The Georgian Bay route was criticized for the number of dams used to store water for its canals. What if one of the dams failed? How would it impact shipping along the route? They also harkened back to Shanly's report of 1858, which cast doubt on the ability to capture enough water at the summit to feed the canals. The strongest argument came from the 1908 report itself, which said that even though the Georgian Bay route was considerably shorter, it may not make for faster passage due to its narrow confines. The St. Lawrence route had Lakes Ontario and Erie for ships to get up a full head of steam.

The next question was who would be paying for the work. The group argued that "Old Ontario," which was assumed to be the settled areas along the Great Lakes and St. Lawrence River, would pay as they had the largest tax base. "Old Ontario" could end up paying for a Georgian Bay Canal that would not improve its own fortunes. Laurier was caught in a flurry of communications from both sides, which meant to choose one side over the other might mean alienating an entire portion of the country. He thus chose to do nothing prior to the election of 1911.

After the election of Robert Borden and the Conservatives in 1911, the pressure on the government continued as the new Erie Canal that bypassed Montreal was nearing completion. The canal supporters were challenged by Adam Beck and the Ontario Hydro-Electric Power Commission, who since inception in 1906, had worked toward the public ownership of hydro-electricity. Beck had met with the members of the Georgian Bay Canal Company and warned the government to make sure that the Georgian Bay Canal was not a gigantic power project in disguise.

In March of 1912, Borden's new government decided to apportion the following sums to railway and canal projects. The Hudson's Bay Railway received $1.5 million, the Welland Ship Canal $200,000, the Lachine Canal on the St. Lawrence $200,000, and $100,000 was allocated to start the Georgian Bay Canal. He had previously tipped his hat in a speech on the eve of the 1911 election: "I strongly advocate the development of our national waterways… Thoroughly equip our Georgian Bay ports, our national waterways, our St. Lawrence route, and our ports on the Atlantic coast."[46] In 1913, he and Robert Rogers, the Minister of Public Works, appropriated $1.66 million for the beginning of the new Welland Canal.

One can assume local disappointment with all sums except that of the Hudson's Bay Railway. History tells us that the work on the Welland Canal was not completed until 1932, and the Hudson's Bay Railway was a disappointment as a grain route from its inception in 1929.

The building of the Hudson's Bay Railway was in part a response by the federal government to quell farmer unrest in the prairies. The farmers believed they were being swindled by the grain elevator companies, including the Northern Elevator Company, the Manitoba Elevator Company, and elevators run by the Ogilvie and Woods Milling companies. These outfits were secretly meeting to fix the price of grain offered to farmers in order to take a larger share of the profits for themselves. The farmers demanded a government-owned railway with similarly owned elevators and got their wish in the form of the Hudson's Bay Railway. The farmers were a tough and powerful lobby with a view "… that all wheat is No. 1 hard, all grain buyers are thieves, and that hell is divided equally between the railways and the milling companies."[47] Meanwhile, in the election year of 1911 money was given by Laurier to finish the last link in Canada's third transcontinental rail line, the thousand-mile link across the shield of Northern Ontario. In practical terms, these decisions seem questionable at best.

Arguments on both sides of the canal issue continued in 1913 and 1914 with groups in favour from Ottawa and Montreal approaching the government and those opposed coming from Toronto and Western Ontario. In the summer of 1914, while enjoying a leisurely holiday at the posh Royal Muskoka Hotel on Lake Rosseau, Sir Robert Borden was recalled to Ottawa at the news of Great Britain declaring war on Germany. All thoughts turned to the war effort.

Despite the looming conflict another commission was appointed by Borden to study the Georgian Bay Canal. This was in addition to the $700,000 spent by the Laurier government on the survey from 1904 to 1908. The new commission was tasked with determining if the Georgian Bay route was worth the $100 million that the 1908 survey had indicated it would cost. It was formed in 1914 and completed in 1917. It was headed by the former mayor of Winnipeg, W. Sanford Evans. In the *Interim Report of the Statistical Examination of Certain General Conditions of Transportation bearing on the Economic Problem of the Proposed Georgian Bay Canal*,[48] the commissioners looked at the current and potential shipping traffic going through Sault Ste. Marie and whether that traffic would support another canal. Several important observations were made. The first was that the two main cargoes through the Soo were iron ore from Minnesota eastbound and coal from Pennsylvania westbound. These were to support steel mills in Gary, Chicago, and elsewhere. Grain was the third largest bulk item.

The grain coming from the U.S. was primarily destined for ports on Lake Erie, of which Buffalo was the largest. Additional American grain was shipped through the St. Lawrence to Montreal. The ships used for these purposes were larger than Canadian ships and thus more profitable. They also had the benefit of more types of return bulk cargo. The Americans were taking full advantage of the shared St. Lawrence Seaway! Canadian return cargo at the time tended to be manufactured goods from Europe and the U.S., which took up much less room on ships. The idea of matching Nova Scotia coal as a return cargo with Lake Superior iron ore was mentioned but not detailed.

The majority of grain produced in Canada was destined for export to European markets. This grain could either be shipped in Canadian or American bottoms, and the main trans-shipment ports were Montreal and New York. At the time, two thirds of Canadian wheat went through American ports. This was despite the fact that the shipping rates through Montreal were slightly cheaper than New York. The customers for Canadian wheat were in order of size of purchase in 1913 Great Britain, France, Germany, and Italy.

The drawback in shipping Canadian wheat by canal was the shortened season due to freeze-up. This allowed the railways to flourish, since they could ship year-round. The railways of the time actually shipped more wheat to ocean ports than freighters. Another shortcoming was the timing of the harvest. In the months of September, October, and November, Canadians sold their wheat to grain companies throughout the world. This uneven yearly distribution was a problem for shipping companies and for customers who wished a more regular flow of product.

The commission's interim report concludes with the argument that ocean shipping rates were subject to the laws of supply and demand and were extremely variable, so much so that it did not matter to shipping companies what savings they could obtain by inland shipping because the ocean rates were far higher. The data used for comparison were ocean shipping rates in 1915, at the start of World War I. The rates had skyrocketed during the war for reasons too obvious to state. In peacetime, the ocean rates were low and relatively stable. The commission also found that most of the shipping traffic on the Great Lakes was coal and iron destined to Great Lakes Ports. The Georgian Bay Canal would not be in effect for this traffic.

The commission used these arguments, weak as some were, to cast strong doubt on the viability of the Georgian Bay Canal. They looked at the canal as simply a through shipping route and not a potential economic stimulus for the towns and cities along it. They overlooked the value of the hydro-electricity generated, again as a stimulus to economic growth. They did not take into account the military benefits of the canal and most importantly did not forecast the growth in the Canadian wheat trade in decades to come. The one most powerful and salient point mentioned was that the canal should already have been built at the time of writing the report, 1917, to take advantage of increasing grain yields in western Canada.

13
THE FRENCH RIVER IMPROVEMENT AND POWER SCHEME

BEGINNING IN 1919, A GROUP OF CITIZENS FROM NORTH BAY, ENDORSED BY THEN MAYOR John McIntyre Ferguson, created a lobby group called the French River Improvement Statistical Committee. The aim of the committee was to convince the federal government to at least build the canal from Georgian Bay as far as North Bay. As early as 1907, the Department of Public Works had initiated dam-building at Chaudière on the French River to control water flow coming out of Lake Nipissing. Nipissing was shallow to begin with, a mean depth of only 4.5 metres, and up to 1.5 metres shallower in the summer months. The impact of lowering the lake on the French River was enormous as a one-centimetre drop in Lake Nipissing led to a forty-eight-centimetre rise in the French River system. Dams were built across the main outlets of the French at Big Chaudière and Little Chaudière in 1907–8. These log dams proved inadequate and were replaced by more substantial structures between 1910 and 1916. One presumes that these measures completed in 1916 were designed to support the Georgian Bay Canal. In a pamphlet produced in 1919, the French River Improvement Statistical Committee outlined their plan for three locks on the French River and a construction time of two years for a canal and three hydro dams at Chaudière, Five Mile Rapids, and Dalles Rapids. The estimated power output from the three dams was 35,394 horsepower.

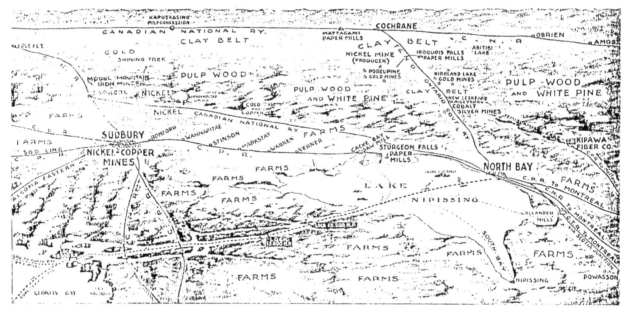

FRENCH RIVER IMPROVEMENT
Give Northern Ontario her Waterway, and she gives "The National" its most economic Grain route.

THE CANAL SYSTEM PROPOSED IN 1919.
(courtesy of Ontario Archives)

Canal cut at Chaudière

The new canal would have a draft of twenty-one feet, which was adequate at the time for the largest lake boats. There was a suitable harbour entrance from Georgian Bay coupled with four deep-water ports on Lake Nipissing at Cache Bay, Callander, Sturgeon Falls, and North Bay. Three rail lines, the CPR, CNR, and Temiskaming and Northern Ontario were poised to take products from the west and Northern Ontario to markets abroad.

The slogan of the committee was "Give Northern Ontario her Waterway and she gives 'the National' its most economic grain route."[49] This was thought to be an added bonus to the plan as the newly nationalized Canadian National Railway was in need of business. In a booklet of some seventy pages, the authors outlined all the ways in which this canal would assist northern development as well as trans-Canada shipping.

They accurately outlined the vast mineral wealth of the north, beginning with Sudbury and its large mines such as those of the International Nickel Company, Mond Nickel, and the British American Nickel Company. These mines were in need of power from the canal to continue their operations. The sale of this power could also help to pay for the waterway. The largest known deposit of silver in the world had been discovered at Cobalt, and the huge Hollinger Gold Mine was in need of power, as well as easy export of its product. Iron had been discovered one hundred miles north of North Bay and was awaiting development. All of these mines needed coal to smelt their ores, and the cheapest way to get it was through the Great Lakes to the French River and inland.

The booklet also detailed the wealth of forest resources in the north and their uses. Available were hardwoods for flooring and furniture, hemlock for rail ties, cedar for telephone and hydro poles, and spruce for large pulp and paper mills such as Abitibi Paper at Iroquois Falls. Finally, it was mentioned that the vast Clay Belt of Northern Ontario was seen as a source of wheat for export.

The familiar argument of using the waterway to avoid shipping through the United States was detailed. According to committee statistics, fifty percent of Canadian grain was being carried by United States vessels, as opposed to only one percent of American wheat being carried by Canadian ships. This route was also forecast to be cheaper than water and rail routes through the United States, both in terms of distance and due to a lower railway grade to Montreal.

The statistical committee of J.B. McDougall, H.M. Anderson, R.A. Tyler, and Cyril T. Young had done their homework and received the endorsement of fifty-seven boards of trade, the principals of the three rail lines, Premier Hearst and his Minister of Lands, Forest, and Mines, Howard Ferguson. A delegation of over one hundred businessmen and politicians ventured to Ottawa on January 10, 1919, to present their case. Once again, the group was met with disappointment as Prime Minister Borden claimed that the government had insufficient funds to finance the project.

All was quiet on the Montreal, Ottawa, and Georgian Bay Canal front until after the First World War and remained relatively quiet until 1925. In 1921 and 1922, the Montreal Board of Trade lobbied the government to reconsider the Georgian Bay Route. They were possibly in favour of this route because some canal boats would have to offload at Montreal onto ocean-going ships. In 1922, a railway strike in the United States caused the clogging of American inland ports and elevators. This situation, it was pointed out, would have been averted with more inland waterways. Charles R. Harrison, MP for Nipissing, asked the government once again to build the section from Georgian Bay to North Bay but was met with an answer from F.B. Carrell, minister of public works, that there was no money available at the time.

Robert Perks, the British investor who still held the charter to build the canal, was fed up with Canadian government inaction and hired Clifford Sifton to sell the charter to the government. In 1923, Prime Minister Mackenzie King told Perks that the government had plans for the Georgian Bay Canal, but it was not presently a top priority. He also told supporters of both the Welland and Georgian Bay schemes that due to the heavy debt of the Canadian National Railway, he was not able to assist either project. Factors in his decision were that King was facing some pressure from the United States to proceed with the deepening of the St. Lawrence route; the costs of construction of the Georgian Bay Canal had escalated since the 1908 report; and there was a brewing dispute between the federal government and the provinces over ownership of water power along navigational routes. In addition, the multitude of Canadian railways feared competition from the canal. Into this fray jumped Clifford Sifton and his sons Harry and Winfield.

14

THE SIFTON PLAN

CLIFFORD SIFTON WAS ONE OF THE MOST CAPABLE POLITICIANS OF HIS TIME. BORN INTO A Wesleyan Methodist family in Arva, Ontario in 1861, his father moved the family to Winnipeg in 1875. Clifford was educated at Victoria College in Cobourg, where he finished first in his class. He returned to Winnipeg to article as a lawyer and set up practice with his brother Arthur in Brandon, Manitoba. In 1888, he entered provincial politics and won the seat as MLA for Brandon under leader Thomas Greenway. Greenway was opposed to the policies of Sir John A. Macdonald and the monopoly of the Canadian Pacific Railway in particular. Greenway fought for greater rights for the province of Manitoba, which came to the fore in the Manitoba Schools Question.

In 1890, Greenway established a "national schools" program in Manitoba in which only provincial schools would be funded. Separate denominational schools, which were funded under the *Manitoba Act* of 1870, would be allowed to exist but would not receive funding. This new arrangement was challenged by the Catholic minority at all levels of justice, including the Privy Council in Britain. Sifton, who was both Attorney General (1891) and Education Minister (1892), successfully argued the case for nondenominational schools and is credited with assisting in drafting the Laurier-Greenway compromise. In it, religious instruction was allowed in national schools, as was bilingual education in a school where a minimum of ten students spoke a language other than English.

Sifton's work on the Manitoba schools question drew him into the sphere of federal politics and the Liberal party under Laurier. In 1896, with the Liberal victory, Sifton joined Laurier and was appointed Minister of the Interior. He was given the task of encouraging immigration to the Canadian west. In this role, his star shone brightly. He and the Immigration Department actively pursued agricultural immigrants from the United States, Britain, and Central Europe. They sought only those individuals with a farming background. He was able to increase the flow of immigrants to Canada from 16,835 in 1896 to 141,465 in 1905. In what may be an insight into the complex personality of the man, he strongly defended the "stalwart peasants in sheepskin coats" from Eastern Europe who were able to quickly establish productive farms on the prairies. In contrast, he largely ignored Indigenous Canadians in his dual role as Superintendent General of Indian Affairs. In this role, he cut administrative costs of the department, as well as funding for native education.

Sifton was a masterful political organizer and strategist who greatly aided the Liberals re-elections in 1900 and 1904. In 1898, he purchased the *Manitoba Free Press* and along with editor James Wesley Dafoe was able to use the popular daily to influence voters west of Ontario.

Parliament was stunned when he resigned from cabinet in 1905, apparently over a disagreement with Laurier on the organization of schools in the new provinces of Alberta and Saskatchewan. Another factor in his decision may have been his lack of promotion by Laurier to more prestigious cabinet positions. Sifton remained in Parliament as an MP until 1911, when he completely severed ties with the Liberal party. He strongly disagreed with the election platform proposed by the Liberals, which favoured reciprocity with the United States. He feared, as did many Canadians, that Canada would be swept up economically and possibly politically by their more dynamic and powerful neighbour to the south.

Sifton went as far as to state his views in a speech to Parliament and to actively campaign against reciprocity in election speeches around the nation. After an anti-reciprocity speech at McGill University in which he and fellow speaker Stephen Leacock were heckled, their carriage, destined for the Windsor Hotel, was overturned by a mob of pro-reciprocity students. The two famous Canadians were dumped onto a muddy street and had to walk to their hotel. Robert Borden, the victorious Conservative leader, admitted that the work of Sifton greatly aided his 1911 election win.

Sifton continued to make contributions to his country through his appointment to the Commission on Conservation from 1909 to 1918. Later, he was appointed to the St. Lawrence Seaway Commission.

He and his wife Elizabeth Arma had four sons, all of whom fought in the war. They moved to Britain during the war years to be closer to their sons. Sifton was instrumental in fundraising for the Canadian Motorized Machine Gun Unit that fought in the war.

Upon returning to Canada, he remained loosely involved in politics as an advisor to Prime Ministers Borden and King. He gave several speeches in support of King's 1925 election bid. He was both highly regarded for his analysis of complex issues and highly mistrusted for his previous defection. There was always a cloud of suspicion over the family as to the source of their wealth. He was investigated for past corruption in his role as Minister of the Interior, but no charges were ever proven.

Clifford Sifton had some previous experience with the canal project, since in 1909 Robert Perks had hired him to lobby the federal government to get the project underway. In exchange, Sifton would receive one third of all future profits of the canal and the hydro power it generated. He resigned from this position in 1910 because it was in conflict with his position on the Conservation Commission. He was also asked by Perks in 1923 to sell the canal charter to the Canadian government. Arthur Meighen, then minister of finance, replied to an inquiry from Sifton, "The fact is the subject of the Georgian Bay canal dropped pretty well out of site during the war, and has not to my knowledge been up for discussion any time since."[50] The answer was a resounding no.

Sifton, since his early days in Manitoba politics, had been strongly opposed to monopoly business practices. In 1921, he began an attack on the Great Lakes shipping interests through the Free Press. "Of no less concern to the western farmer was the manipulation of rates for grain, flour, and cattle by combinations of shippers on the Great Lakes and the Atlantic Ocean."[51] Sifton had always contended that a Canadian national transportation policy should encourage the shipment of goods from Canadian ports in Canadian ships. He believed that transportation companies, both Canadian and American, were conspiring to avoid competition and skim profits from the farm producers. The ultimate result was damaging to Canada's national interests. He was suspicious of the situation in 1920, but it was not until September 1922 that the *Free Press* unleashed a steady and detailed assault that led directly to government commissions to investigate the transportation monopolies. As part of the agitation, Sifton himself wrote a series of front-page articles for the *Free Press* in a racy, trenchant style. This was the only time he became directly involved in writing for the paper.

Of equal concern to Sifton were the mergers in the Canadian banking industry and the potential for monopoly practices. When the Royal Bank swallowed up the Union Bank in 1925, he expressed his concern that the "Bank of Montreal crowd and [Herbert] Holt crowd were slowly acquiring a stranglehold on the business community, making it more and more difficult for anyone out of favour with the small clique who controlled the banks to acquire substantial investment capital."[52] Sifton had few kind words for Herbert Holt, possibly Canada's richest man at the time. He had previously sought support for his mobile machine gun unit from Holt, who gave him a bit of cash and told him he knew of a motorcycle in Britain that he could have. Holt was in control of much of Montreal's business life through his Montreal Light and Power Company, itself a monopoly. The two would do battle indirectly over the Georgian Bay canal in the next few years.

In 1924, Winfield Sifton, Clifford's son, developed a plan that he believed would finally get the canal project underway. Winfield was the second of Clifford and Elizabeth's five sons. Winfield was called to the bar in 1913 but chose a business career over a law career. He dabbled in the import export trade, marine insurance and securities, amongst other businesses. He fought alongside his brothers in the war and remained in England for a time, promoting Sifton business interests. A friend of the family commented, "He seems, however, to have inherited his grandfather's propensity for spreading his energies in many areas at once and ultimately failing in most things he tried."[53]

The third son, Harry, also became involved in the canal project as his father's personal secretary from 1919 until his death in 1929. Harry was easygoing and the most popular of Clifford's four sons. He was a horseman of dominion fame, a director of newspaper companies, and later a land developer in London, Ontario.

In 1924 and 1925, Winfield ruminated on a plan. He also had the foresight to have a plan of the route submitted to the Department of Railways and Canals in 1925. This was a hastily conceived plan, and apparently he was aware of this lack of oversight as he also applied for and received three rulings from the Department of Justice. The first was to say that not all parts of the plan needed approval before construction began. Secondly, if the company failed to complete the work on the canal by 1933, they would still retain the rights to develop water power. Third, if the company submitted detailed plans to the government that were approved, then the company could neglect the canal and proceed with the water power. Clearly, the Siftons were thinking ahead and had their eye on securing water power for the canal company.

Winfield was under some time pressure as the renewal of the company charter was due in 1927, and he had competition for a portion of the canal. In 1921, the National Hydro Electric Company had obtained a charter to build a power plant on the Ottawa River at Carillon. Robert Perks, who owned the Georgian Bay Canal charter until 1926, objected to the plans at Carillon because his company's charter, first approved in 1894, gave them the right to develop power along the Ottawa River.

In the newly hatched Sifton plan, the ever-expanding market for electricity in Ontario would provide the majority of the revenues necessary to complete the canal, with the remainder to be raised in the private market. Simply by developing all the hydro-power sites, including Carillon, as the canal was built, the project would pay for itself. The government would thus be under no further financial obligation. All that was necessary was permission to proceed. Another element of the plan was a clause that at any time during construction of the canal, the federal government could take it over with short notice by only reimbursing the canal company their expenditures, to that date. Also, the company would be allowed to charge a $1 per ton toll on shipping on the canal for ten years to recoup construction costs. Finally, power developed along the canal would be sold at a price determined by the federal government.

In April of 1926, the final details of the Sifton plan for the canal were published. Essentially, the route remained the same as outlined in the 1908 Public Works Report with the exception that canals had been deepened to twenty-five feet to carry ocean-going vessels up to 15,000 tons. The costs had escalated significantly since 1908. The estimate by company engineer G.W. Volkman was $197 million for the canal and $93 million to develop 875,000 horsepower of electricity along the route. This was to be financed entirely by private Canadian and British investors and the investment repaid by charging a toll of $1 per ton on ships using the route. The toll was reasonable by comparison, since the toll on the Suez Canal was $1.50 and the Panama Canal $.90. It was, however, a new wrinkle to shipping in Canada, since previously no direct tolls had been charged. As power was developed along the route and sold, the toll could be reduced or eliminated. There were no government subsidies, land grants, or bond guarantees requested, which was somewhat unique in the building of large public works throughout Canadian history.

The canal company believed that shipping goods from the Canadian and American northwest via this route was more cost-effective. It would save approximately twenty-two hours over the St. Lawrence Waterway and an additional twenty-one hours over the All American or Erie Canal route. In terms of distance, it claimed to be 282 miles (453.7 km.) shorter from Fort William to Montreal than the St. Lawrence and 766 miles (1232 km.) shorter than the Erie track. A savings of four cents per bushel of grain were claimed. In terms of the North American shipping industry, this all-Canadian route would keep Canadian goods in Canadian bottoms and divert some U.S. shipping into Canadian waters. It was believed that if constructed, the U.S. would build its Lake Champlain Canal to the Hudson River, requiring American ships to pass Montreal on their way to New York. It was believed that eighteen U.S. states could take advantage of the Georgian Bay Canal. The plan also claimed that shipping by water was nine times cheaper than shipping by rail.

Winfield had made the acquaintance of Perks and the other British investors during the war, and they were quite willing to let him have control of the canal project if he could succeed. Robert Perks sold the charter to a company controlled by Sifton's sons, the Great Lakes Securities Company, which in 1926 sold it to a holding company controlled by Clifford. In November 1925, as these plans were taking shape, the Siftons interviewed Prime Minister King three times.

King had said of Clifford that he had few if any friends left after abandoning the Liberal Party in 1911. He was also critical of Clifford Sifton's wealth. On visiting the Sifton family estate in Toronto (Armadale), he remarked, "I noticed a huge vault on the way (through the house). Is it all indicative of his power? He has amassed great wealth somehow and is a sort of medieval Baron."[54]

Winfield took up the task of trying to lobby King. He assured King that he had reached an agreement with Ontario Hydro to take the electricity generated by the canal and that no power would be exported from the canal route sites to the United States. King replied that if the government were to support the scheme, Sir Clifford should guarantee that the *Manitoba Free Press* would, in effect, be a government supporter. The Siftons would go further than that, responded Winfield. They would spend up to $2 million to acquire control of the *Globe* to ensure its support of the King government. On the one occasion that King saw Sir Clifford himself, the latter informed the Prime Minister, perhaps somewhat disingenuously, that he was not behind the project for the money. He had all he needed, and his sons had all that was good for them. Instead, this was a project essential to Canada's development that would involve the expenditure of many millions of dollars in Canada at no charge to the government. King was sympathetic to the plan, considering the political shoals that he knew lay ahead.

One of the first to object to the Sifton plan was Ontario premier Howard Ferguson. He did not believe that hydro power resources on the route should be handed over to private interests. He threatened legal action if the plan were to go through. On March 8, 1927, he hastily introduced a resolution in the Ontario Legislature in opposition to the renewal of the canal company charter.[55] In response to a question from a

Star reporter on why he objected to the plan, Ferguson replied, "Power, of course. To renew the Georgian Bay Canal charter is simply to hand over to a private group a ready-made canal in the Ottawa River with enough power to run the canal part and to make them wealthy. We must keep it. Ontario needs it for herself. The canal charter is just a power charter."[56] Harry Sifton responded in a newspaper interview by pointing a finger at Ferguson and the Quebec Power Trust headed by Herbert Holt. Agreements with Ontario Hydro to develop water power at Carillon and Gatineau and sell it cheaply to the Ontario Hydro Commission were being discussed.

Herbert Holt

An aside on Herbert Holt. His death in 1941 was announced at a triple A baseball game in his hometown of Montreal. The crowd cheered the news. He was considered one of '*les maudit anglais*', the Anglo-Protestant power elite of Montreal that controlled business in Quebec and beyond.

Born in Ireland, Holt emigrated to Canada in 1873 and went to work as a civil engineer. He spent the first twenty years of his life as an engineer, first with the Toronto Water Works and then the CPR.

While surveying and building the railway, he worked with the likes of Mann, Mackenzie, and Thomas Shaugnessy. Later, he moved to Montreal and focussed his considerable energies on acquiring companies. His first venture was the Montreal Light, Heat and Power Company, with which he consolidated the gas and electricity production and distribution in Canada's largest city. He went on to become president of twenty-seven companies and on the board of directors of three hundred. He was chairman of the Royal Bank of Canada from 1908 to 1934 and listed in his portfolio of companies mining, railroads, streetcars, forestry, flour mills, shipyards, movie theatres, life insurance, hydro and gas, and more. One Montreal resident in the late 1920s remarked, "We get up in the morning and switch on one of Holt's lights, cook breakfast on Holt's gas, smoke one of Holt's cigarettes, read the morning news printed on Holt's paper, ride to work in one of Holt's streetcars, sit in an office heated by Holt's coal, then at night go to a film in one of Holt's theatres."[57] At the height of his success, he controlled assets worth over $3 billion. Margaret Westley, who authored *The Anglo Protestant Elite of Montreal*, said, "Everyone respected his business ability but no one liked him personally."[58] Holt tended, as did many businessmen of the time, to try to gain complete control of certain markets, in other words to gain a monopoly on all aspects of the production, distribution, and consumption of certain essential goods and services. His business interests were intertwined with the business life of Quebec.

Holt did know how to take care of his political friends. Beginning in 1922, he, along with recently appointed high commissioner to London, Peter Larkin, and Senator Arthur Hardy, established a personal fund for Prime Minister King to supplement his modest $10,000 salary. By 1930, this fund had reached an impressive $237,000.

In the mid-1920s, the ownership of National Hydro Electric Company fell under the umbrella of the Shawinigan Power Company and the powerful Herbert Holt. A contest for control of power rights on the Ottawa River was begun.

15
POWER PLAYS

THE BACKGROUND TO THE EMERGING CARILLON CONTROVERSY IS INSTRUCTIVE AS IT BECAME a direct competitor of the Georgian Bay Canal Company and their plans. In 1909, a small power lease was granted to the Carillon Power Company to develop 250 horsepower of electricity at Carillon in association with the Carillon canal. The principles in this deal were the National Hydro Electric Company headed by Messrs. Miles, Gosselin, and Robert. In December of 1921, prior to the federal election, the Meighen government granted a larger lease to NHEC. This new lease would allow NHEC to build a new dam and drown existing canals. The company would build a new canal to a depth of nine feet, nowhere near deep enough for lake freighters. This lease allowed the company to take control of power at the site until 2006! No work was done on the lease between 1921 and 1926. The NHEC could not raise the capital for the project, so the lease was sold to the Shawinigan Power Company, one of the players in the Quebec Power Trust, which included Herbert Holt. The lease was again renewed in 1926, with even more favourable terms for the lessee. The lease was for 300,000 horsepower of electricity, of which 100,000 horsepower was designated to be shared with Boston industrialists. Before arrangements could be finalized in 1926, Quebec premier Taschereau publicly stated that no power from his province would subsequently be exported to the United States. This was a reversal of protocol for the premier, who had actively sought American, Brazilian, and British money to develop power and industry in his province. Taschereau was believed to be concerned with Prime Minister King's plans for the St. Lawrence Seaway, which included selling power to the United States. He may have also objected on grounds that the power sold would aid industries and employment in the United States and not Quebec.

The corporate entity behind the lease was the Holt-Shawinigan group. Although two separate firms, they did control a number of companies through their co-ownership of United Securities Limited. They also shared the same office building in downtown Montreal. John Edward Aldred, the American head of the Shawinigan group, sat on the board of directors of Holt's Montreal Light Heat and Power Company, as did Holt on Shawinigan Power. On October 5, 1926, the *Toronto Star* reported that Ontario and Quebec had reached an agreement with the NHEC and Shawinigan Power to develop and build a hydro dam at Carillon of some 133 feet in height and producing between 250,000 and 300,000 horsepower of electricity. Provision would also be made for a navigation channel between Carillon and Montreal of fourteen feet in depth. It was to be approved by the Quebec legislature the following week. Ontario Premier Ferguson and Hydro chairman Magrath were said to be delighted on the potential approval. The work was to be done by the NHEC. In return for the rights to this power, the NHEC would pay the government of Quebec an annual rental of $25,000 for seventy-five years plus fifty cents per horsepower of electricity generated. In the terms of the lease, power rentals would not start until 1932, and if NHEC sold power to the provinces of Quebec and Ontario, it would receive two thirds of the lease payments back. This amounted to a $600,000 savings compared to the 1921 lease. The financing of the project was to come from future contracts for power with the Shawinigan group, which had built power plants at Shawinigan Falls and elsewhere on the St. Maurice River, as well as the Ontario Hydro Commission. In the 1926 deal, the beneficiaries were to be the parent company for providing much needed power particularly to Ontario. This deal seemed too good to be true. Could this be a case of collusion prompted by political patronage? The two provincial premiers involved in the deal were Howard Ferguson and Louis Taschereau.

Laurel and Hardy

The two premiers could not have differed more in appearance. Ferguson was a fleshy small-town Ontario Orangeman with a bulbous nose. Taschereau was a slightly built, fine-featured French Canadian from one of the first families of Quebec. Louis-Alexandre's father had been a judge of the Quebec Supreme Court. Together they resembled the famous comedy duo, Laurel and Hardy. Their combined efforts to secure the rights to hydro-electricity for Canadian provinces were no laughing matter. These endeavours would shape the future of Canadian resource development.

"Foxy" Fergie had been the Minister of Lands, Forests, and Mines prior to becoming the Conservative leader, then Premier of Ontario in 1923. In his role with Lands and Forests, he was seen to have entered the government of Ontario into several questionable speculative pulp contracts in the Lakehead region. He granted huge tracts of land to timber licensees, foregoing the typical auction format and assigning the licences on an arbitrary basis. He was called the most corrupt influence in the provincial government of William Hurst. When the Drury government (United Farmers) took power from the Conservatives in 1919, they investigated Fergie's actions as minister in an inquiry called the Timber Commission. Ferguson was found to be personal friends with lumber barons E.W. Backus and Jim Mathieu, whose companies had made a mockery of existing timber regulations. As an example, their firms would hire government cullers to work for them when they were not estimating the volume and value of timber being cut for the government. Foxy Fergie was able to ride out the tide of criticism by turning the tables on the commission judges, Latchford and Riddell. He successfully insinuated that they had it out for him all along and that the commission was little but a witch hunt. He subsequently uncovered evidence that several of the Drury government ministers were involved in similar quasi-legal acts.

Louis Taschereau was the premier of Quebec from 1920 to 1936. He aggressively promoted the development of Quebec's resources, including hydro power. A goal of his administration was to stop the exodus of Quebec workers to the industrialized states of the northwest United States by creating industrial jobs at home. He sought capital for large hydro projects from American interests. "We want to bring in new industries and we are ready to do all that is possible in that direction. We are not afraid of foreign capital… I prefer importing American dollars to exporting Canadian workmen."[59] Among his allies was the American head of Shawinigan Power, J.E. Aldred, of whom he said had shown "a big heart, a clear vision of the future and an abominable will to succeed."[60] After four terms in power, Taschereau's star began to fade. Rates for electricity charged by Montreal Power, Heat and Light, Shawinigan Power, and Quebec Power were seen to be exorbitant in depression times. It did not help Taschereau that he had a brother, Edmond, who sat on the board of Quebec Power and another brother, Charles, who worked for Shawinigan. In the despair of the thirties, he was blamed for allowing men such as Herbert Holt to become fabulously wealthy by monopoly control of Quebec's natural resources. He was investigated by future premier Maurice Duplessis for unauthorized use of public funds and patronage to family members and friends. The most egregious of these acts was putting his older and less astute brother, Antoine, in custody of government members' salaries and expenses. Antoine took this money, invested it, and reaped the interest generated. Taschereau was defeated in 1936 by the Union Nationale party led by Duplessis.

Howard Ferguson and the then chairman of Ontario Hydro, Charles Alexander Magrath, were intent on buying power from the Holt Shawinigan group at Carillon and developing their own hydro facility at Chats Falls on the Ottawa River. Ontario was in need of hydro for the smelting of ores, the pulp-and-paper industry, and the emerging chemical industry in the province, as well as for residential customers. They needed to service a 400% growth in hydro usage in Ontario from 1921 to 1930.

The break for both provinces came when the International Paper Company negotiated with Gatineau Power to buy power from the Quebec firm and share up to 260,000 horsepower with Ontario Hydro. Magrath told a *Star* reporter that he would rather buy power from private interests in Quebec than the Georgian Bay Canal Company, whose principal shareholders were largely from Ontario. A *Star* reporter commented, "Unfortunately the compliments paid to Mr. Ferguson for dealing with an emergency (the need for power in Ontario) seem to have encouraged him to make new and more objectionable arrangements with power interests in Quebec. The premier seems to be obsessed with the idea that it is good business to let private parties put up the money for power plants, seemingly overlooking the fact that in the end the people of Ontario will pay in their rates for the plants and have not one steel girder of their own to show for their outlay. Under such conditions the people will not get service at cost, but will contribute to the fortunes of private investors."[61]

Magrath's chief concern was an impending legal battle between the provinces and the federal government on the control of power resources, which would seriously delay developments on the Ottawa River. In a letter to the mayor of Pembroke in response to a request that someone in his department attend an information session on the canal project, Magrath replied, "The whole idea of handing over to private interests an international waterway of over two hundred miles in length, in these enlightened days, is repugnant to all sane national views."[62]

The two premiers collectively fought to retain control of the banks and beds of waterways in their provinces. In a letter to Taschereau on the canal charter, Ferguson wrote, "My understanding is that we are both determined to resist to the very utmost any movement of that kind." Ferguson's stance was that Ontario and Quebec should cooperatively develop power on the Ottawa River and with federal help make some improvements in navigation.

Local disappointment with the support of the Carillon deal and the resistance to the canal charter by the provincial premiers was swift in coming. On October 11, 1926, Nathan Couchon, chairman of the Ottawa planning commission, told the *Star* that the city of Ottawa was anticipating the building of the Georgian Bay Canal and that no work at Carillon should be undertaken without a guarantee of a 30' deep lock between Carillon and Montreal. On December 4, the *Star* reported that the arrangement between the provinces and the federal government at Carillon, signed by the then Minister of Railways and Canals, had in effect given the provinces the right to develop power at Carillon. The incoming Minister of Railways and Canals, C.A. Dunning, held that the rights to navigation on the Ottawa took precedence over the provinces rights to power.[63]

> *A Star reporter attempts to add clarity to the situation. The fight for possession of the Carillon water rights still goes on merrily at Ottawa between the holders of the Georgian Bay Ship Canal charter and the holders of the water rights of the National Hydro Electric Company. Both rights are about to expire, unless renewed. The present situation has developed because the bill to renew the rights for the Georgian Bay Ship Canal has come up first for consideration.*

The Ship Canal supporters say that if the nation is not prepared to build the canal along the Ottawa River their company should be allowed to proceed. They think they should be permitted to finance the work, in part, by developing power and charging tolls to shipping. And they claim that their existing water rights take precedence over those of the National Hydro Electric Company.

The Ontario Hydro Commission is interested because the governments of Ontario and Quebec have agreed upon a fifty-fifty division of the water available for power at Carillon and the commission is negotiating with the National Hydro Electric Company for a joint development in Carillon. These negotiations have not come to a head and there are rumors of a hitch. The Holt and Shawinigan interests have got behind the National Hydro Electric Company. It is being suggested that the Quebec power ring would not care to enter into a partnership with Ontario's public ownership trust and have power sold at cost in this province while similar power is being sold at a profit in Quebec. On the other hand, it is argued, here and there, that a public ownership enterprise should not co-operate with a determined opponent of the public ownership principle.

There were three issues that the renewal of the canal charter and the Sifton plan raised. The first was who controlled the power produced by dams on the Ottawa River or anywhere else on the route. The federal government argued that if they were building a dam for navigation purposes and in so doing built a head of water, then it was a natural conclusion that they could produce power from that situation, as could any firm that obtained a charter for developing navigation structures along the river. The Ottawa River, according to the Siftons, had been declared a navigable waterway by an Act of Parliament in 1870, and therefore the federal government had control of the power produced. The provinces argued that the riverbed and the resources within it belonged to them, as did the power produced by the water over which it flowed.

The second issue was who controlled the export of power. The Cedar Rapids Power Company, with a hydro dam at the narrowing of the St. Lawrence River near Montreal, and the Ontario Hydro Commission, were involved in the export of power to the United States under a federal license. This situation was challenged by the provinces, but the right of export was held to be under federal jurisdiction. The precedent had been set in a case involving the Trent Valley Canal that the export of power was under the jurisdiction of the federal department of Trade and Commerce. The third was jurisdiction. Not only did you have the Trade and Commerce Department involved, but also the Ministry of Railways and Canals and the Ministry of Public Works. Throw into the mix the provinces of Ontario and Quebec, the Ontario Hydro Commission, and the "Quebec Power Trust," and the situation became extremely complex.

INVESTIGATION OF THE CARILLON LEASE

The 1926 lease at Carillon turns out to have been a behind-the-scenes arrangement that no government officials in either the Department of Railways and Canals or the Ministry of Public Works seem to have known about. On August 20, 1926, Arthur Meighen, then Prime Minister, wired Sir Henry Drayton, the Minister of Railways and Canals, to extend the now Shawinigan Power lease at Carillon. Drayton then wrote Ferguson to say that he would not allow construction to commence until Shawinigan was willing to sell power to Ontario Hydro at satisfactory terms. This sweetheart deal for the two provinces was cancelled by Prime Minister King when back in power in November of 1926 by an order in council. King claimed the Shawinigan group had made contributions to Tory campaign funds.

At a future investigation of this lease and the canal charter, Major Bell (Deputy Minister of Public Works) claimed that he was out of town when the lease was struck and that no one from his department knew anything about the matter. When pressed further in the House of Commons, Bell almost let the cat out of the bag but was prevented from doing so by repeated diversionary questions from Conservative MPs. One member of the house declared, "You had better ask Sir Henry Drayton, he knows all about it."[64] Sir Henry had been the Commissioner of Railways and Canals from 1912 and a former member of the Toronto Power Commission. He was a Conservative member for Kingston from 1919 to 1928, the Minister of Finance in the Borden government from 1919 to 1921, and the Minister of Railways and Canals for a short time in the summer of 1926. When King cancelled the 1926 lease, NHEC returned to the terms of the 1921 lease, which was set to expire on May 1, 1927, the same day as the canal charter.

The supporters of the Carillon Power project including Ferguson and Taschereau, had been silenced … for a time.

The Quebec power trust involving Holt's NHEC and Shawinigan Power objected to the Sifton plan because they wanted to develop the power at Carillon and sell it initially to U.S. concerns. The tentacles of this alliance had connections to American investors who were interested in cheap power and in pulp and paper companies such as International Paper and E.B. Eddy.

The railways and Sir Edward Thornton, president of the Canadian National Railway, objected to the canal, presumably because the canal would create more competition for the struggling railway. Thornton was trying to make a go of combining two of Canada's transcontinental railways, the CNoR and the National Transcontinental, both of which had struggled financially before government takeover.

16
BILL NO. 78

THE BILL (NO. 78) TO RENEW THE CHARTER OF THE MONTREAL, OTTAWA, AND GEORGIAN Bay Canal Company was put forward as a private member's bill by the highly respected E.E. Chevrier of Ottawa. It was the last opportunity for the Siftons to hatch their plan to initiate the building of the Georgian Bay Canal. It was the thirteenth attempt over thirty-three years at renewing the charter; the twelve previous had been successful.

Charles Dunning was intent on killing the bill by having his own Ministry of Railways and Canals staff knock it in the head with unfriendly evidence. The bill passed easily in first reading but was hotly debated in the second reading beginning on February 25, 1927. It was debated twelve times. In many of those instances, the one-hour limit on the debate of private members' bills was surpassed. A somewhat perturbed Conservative member, Archibald M. Carmichael, stated in the House:

> *We had bill No. 78, an Act respecting the Montreal, Ottawa and Georgian Bay Canal Company introduced the 25th of February... It was discussed on March 1 for an hour, again on March 4 for one hour and on Monday March 7, it took practically the whole day. On Tuesday, March 8 it was again discussed for an hour. On Monday March 11 it was again discussed for an hour and on Monday March 14, another day was spent on it. On Tuesday March 15, and on Friday March 18 another hour was devoted to it. On Monday March 21 it was taken up again. On Tuesday March 22, another hour was occupied. On Friday March 25 and again on Monday March 28, another hour was consumed with this bill.*[65]

The sitting members passed the tedium of the debate by making up and singing songs about the proponents of both the Georgian Bay Canal and Carillon deals. The house scribes chose not to include these in Hansard. Carmichael goes on to add that Parliament had reached its limit for discussion of private bills, with Bill No. 78 alone resulting in the delay of other legislation.

A summary of arguments for and against the extension of the charter heard in the House of Commons during the debate of Bill No. 78 begins with Sir George Perly. He was one of the first to speak, calling the charter a hardy perennial that had occupied the time of legislators for thirty-two years. He thought that after such a long time without any actual construction, perhaps the projects time had passed. Sir George was the MP for Argenteuil, which contained the Carillon site. His attitude was that he did not much care who put the power dam in place but to hurry up and get the job done. He also pointed to the under capacity of the Canadian National rail system and the already clogged Port of Montreal as factors against building the canal. He stated that currently in Canada there were no tolls on canals, even on American vessels, so a precedent would be set with the Georgian Bay Canal. A week later, he called for the federal government to take over the project and to reimburse the canal company.

Opposition from other members of the house came in part from a request by the Ontario government that the bill be quashed. The main beef was that the province had developed a publicly owned utility in the Ontario Hydro Commission and did not want to cede power to private interests. Adam Beck, the dynamic founder of Ontario Hydro, passed away in 1925 and was replaced by C.A. Magrath. Magrath, as previously mentioned, balked at the private development of power on Ontario's rivers. The Siftons had anticipated this wrinkle and had previously modified their charter with the help of Adam Beck to read that the charter granted rights to "produce, lease and supply, or otherwise dispose of surplus hydraulic, electric and other kinds of power, the rates to be fixed by the Board of Railway Commissioners."[66] They had also written into the charter that the tolls on the canal could be determined by the same body.

Lapierre, the Liberal member from Nipissing, spoke at some length on the importance of the canal to Northern Ontario. He claimed to be representing the communities of North Bay, Cochrane, Englehart, Renfrew, Mattawa, Pembroke, Parry Sound, and Callander. He believed the industrial development of the North depended on the canal. He cited ores at Haileybury, New Liskeard, and Ville Marie that needed access to coal to be smelted. He bemoaned the fact that International Nickel had moved its smelters far from the Sudbury site of its ores to Port Colborne to take advantage of cheap American coal. Northern Ontario had yet to be serviced by Ontario Hydro. They were seeking a cheaper alternative to the high private company rates they were paying for electricity. He also cited the British American Nickel Company that had moved its smelting operations to Ottawa to have access to water power. He reviewed the many promises from Laurier through Ferguson to build the canal that had not materialized, thereby dashing the hopes of northern Ontario businessmen. Lapierre cited a previous discussion of the plan in the House on February 14, 1910. During the debate, Colonel Arthurs from Parry Sound made a strong case in support of Northern Ontario. He outlined the long history of the charter and stated, "We believe that under this charter we can put through a canal that will not only be a cheaper method for carrying our western products eastward, but will be a great transportation facility in any direction."[67] In his historical outline, Arthurs mentioned a huge rally in support of the bill in 1908 involving almost every constituency in Ontario and Quebec. He also pointed out that the charter of the canal company only allowed the use of power generated to operate the canals. Any additional power generated and sold must have the permission of the lieutenant governors of Ontario and Quebec. Arthurs stated that the western provinces had been in favour of the scheme for decades as a way to reduce shipping costs on grain. They viewed this as a national waterway to get western products to market and manufactures to the western provinces.

Future Prime Minister R.B. Bennett threw his hat into the fray by arguing against the renewal of the charter. The Conservative member from Alberta began by accusing Mackenzie King of meeting with the Siftons in Atlantic City to discuss the charter. This King flatly denied and asked for and received an apology from Bennett, who went on to argue that the bed of the Ottawa River belonged to the provinces, and to build a dam on top of it would require the permission of the provinces. He believed the power generated from rivers belonged to the provinces and should remain with them. The renewal of the charter would override all other leases on the Ottawa River, and thus the hydro power generated would pass out of control of the citizens of Ontario and Quebec.

Bennett was joined in his criticism by DeWitt Carter of Port Colborne on the Welland route. Carter echoed many previous objections to the Georgian Bay Canal. He believed there were too many locks on the Georgian Bay route and that the Georgian Bay coast was too dangerous. He believed there to be rocks extending offshore from the entrance to the river some four miles to the Bustard Islands. He termed the French River "a rocky and broken country" with many sharp bends. He added that he believed Lake Nipissing to be too shallow and that more water was needed at the summit to feed the canals.

Several MPs pointed to the fact that the charter had been in existence for thirty-three years and nothing had been accomplished. They failed to realize that various attempts to get the project underway were denied or overlooked by the governments of Laurier, Meighen, and Borden.

The issue of the government claiming that plans for the canal project were never submitted for approval was raised. In fact, plans *were* submitted, but no one from the government ever responded to those plans. The canal company had reorganized after the purchase of the charter from Robert Perks. The chief shareholder was the Great Lakes Securities Corporation controlled by Clifford Sifton. There were ten additional shareholders, including Winfield Sifton and the engineer, T.W. Volkman. Volkman had submitted a revised set of plans for the work to be done to the Ministry of Railways and Canals. In the House, Colonel Dubuc of the Ministry of Public Works strongly criticized the plans as not detailed enough to allow construction to proceed. He admitted to receiving plans, to forwarding his objections to the plans to his supervisor, the deputy minister, but not communicating in writing with the canal company. His objections included the serpentine nature of the French River, flooding caused by raising water levels on Lake Nipissing and the Ottawa River, and the inadequacy of the plans submitted. In turn, the company called on engineer Coutlee from the 1908 survey, who claimed that the company plans were feasible. The plans of the Department of Public Works from the 1904 to 1908 survey would have been a sufficient starting point for any construction project. Clearly, this was another example of the canal plan being stonewalled.

Two additional concerns raised by Lennox (Conservative, North York) were that if the charter were again approved, the provinces of Ontario and Quebec would sue the federal government, and the water power potential of the Ottawa River would be idled by legislation. Secondly, he thought that the canal company could proceed with developing water power without building the canal. If this occurred, the federal government would have an unfinished navigation canal on its hands while the canal company controlled the water power. By 1927, the fight had become not about how to get goods to and from Canada cheaply but rather who would control the vast water power resources primarily of the Ottawa River.

Hugh Guthrie (Conservative, Guelph) made the argument that the private member's bill should be moved to a public bill. He correctly claimed that the bill put private interests in competition with the national transportation system, i.e., the Canadian National Railway. He also saw the conflict between the federal government rights to navigation being pitted against the right of the provinces to natural resources, including hydro power. His motion to move the bill to public discussion was narrowly defeated.

The private member's bill was passed on from second reading to the Standing Committee on Railways, Canals, and Telephone Lines. Bill No. 78, *An Act Respecting the Montreal, Ottawa, and Georgian Bay Canal Company*, failed to get past second reading by a narrow margin. The next and last review was done by the Commission on Railways and Canals.

17

THE RAILWAY COMMISSION HEARING

ONCE AGAIN, THE SIFTONS AND THEIR ATTORNEY, J.A. RITCHIE, PITCHED THEIR PROPOSAL. Harry began by addressing the concerns expressed by MPs during the second reading of the bill. He said that the spirit of the charter was to create a national waterway and not simply to do a hydro power grab. He stated that no profits would be taken by shareholders until the canal was complete. The majority of directors of the board of the canal company were to be naturalized Canadians. There would be no export of power to the United States. He also stated that the company would complete its plan for construction of the canal, since the plans submitted by the company in 1924 were not complete. He reiterated that the government could at any time take over the canal by paying the company its expenses to the date of the takeover.

Harry favoured the federal development of navigation and power on both the Ottawa and St. Lawrence route. His point was that it was not equitable for the feds to pay for canalization and the provinces reap the rewards of that canalization in the power generated.

Winfield Sifton was the next to testify. He reviewed the situation on the Gatineau River, where a private firm from Quebec (from the Quebec Power Trust) had built a dam and entered into an agreement to sell power on a thirty-year lease to the Ontario Hydro Commission. After thirty years, the private firm would own the power works and would be able to sell power for whatever price it wished. In essence, the taxpayers of Ontario were paying for this private venture. He said his canal company would sell power to Ontario Hydro subject to a clause in the charter that stated they could only charge what the Railway Commission would allow. He also reiterated the relationship between building dams and creating power. "No dam can be built for power which does not affect navigation, and no dam can be built for navigation that does not create power."[68] He outlined that it was common practice for the Department of Railways and Canals to lease power created by navigation dams. His company would do the same at rates set by the government.

The last to speak in favour of the canal was J.A. Ritchie, counsel for the Siftons. Here is a summary of his arguments:[69]

- The canal would be built by private enterprise and would not cost the government a cent.
- Hundreds of millions of dollars would be spent in Canada on its construction.
- It would be a competitive enterprise and not contribute to an existing monopoly.
- It would improve the shipping business in Montreal, Quebec City, Trois-Rivières, Sorel, etc.
- It would reduce the cost of transporting wheat from the Canadian prairies by as much as five cents a bushel.
- It would create many inland harbours along the route.
- It would allow coal to travel inland to markets and northern ores to be transported to Sydney for smelting.
- It would eliminate Buffalo as a port where Canadian products were bled off to American ports.
- It would make large amounts of hydro-electricity available at low cost.
- It would substantially contribute to the industrial development of Ontario and Quebec.

Mr. Ritchie also addressed the animosity that had been shown in the House of Commons and newspapers toward the plan. He talked about those who in 1894 had first sponsored the charter and the fine character of men they were.[70] The point was that the intentions of the founders of the plan had always been honourable and in the best interests of the nation. In a humorous moment in the hearing, Ritchie was asked for the addresses of the men he had described to the committee, most of whom were deceased.

His reply was, "Well, I cannot tell you of the exact addresses of those who are deceased, but you may be sure it is either up or down."[71]

At the conclusion of the hearings, the chair, Charles Dunning, asked one final time for objections to the renewal of the charter to be stated. No one spoke. The matter was put to vote and the renewal of the charter defeated. Dunning had come up with another plan for the Carillon dam, which denied both applicants for the renewal of their request. He suggested that the federal government take over the works at Carillon both for navigation and power and that both the canal charter and the Carillon lease be allowed to lapse.

In response to the Railway and Canal Committee decision, Clifford Sifton, who called Dunning a friend, said, "I have not heretofore known of a minister of the crown making as complete a skunk of himself as Dunning has done in this case."[72] Harry and Winfield Sifton were eager to continue the fight for a canal charter, but Clifford said no to any further effort. The decision had left the boys feeling as if they had been butchered by a plan devised by Dunning. Sir Robert Perks wanted Clifford to sue the federal government for monies spent on the charter over the years, but again he said no. Perks and the New Dominion Syndicate spent the next nine years trying to get compensation for their charter, to no known reward. The Holt/Shawinigan group turned its attention to power projects on the St. Lawrence route. In an ironic yet apparently planned move, Prime Minister King appointed Clifford Sifton to the National Advisory Board for the St. Lawrence Seaway project. He remained in this post until his death in New York in 1929.

Unfortunately for Clifford Sifton, the Canadian public and his fellow politicians were suspicious of his motives in thinking he was advancing the cause of the canal solely for financial gain. Sifton's biographer, David J. Hall, believes he was motivated by a sincere desire to make improvements to the Canadian transportation system in building the Georgian Bay Canal.

The issue of control of Canada's rivers for the creation of hydro power came up again at the Dominion Provincial Conference in November of 1927. The dispute over the canal charter and the Carillon lease were one in a number of disagreements between the provinces and the federal government over the control of hydro power on navigable rivers. The *BNA Act*, section 108, stated that public works and associated property shall be the property of the Dominion of Canada. This included canals and the associated water power. The provinces in the Act were given control of all property rights within their borders. Judicial decisions in the 1920s such as the case of the *Ontario Rivers and Streams Act* favoured provincial control of water power. Ontario in particular objected to allowing surplus power from canals to be given to the federal government. It was willing only to allow the use of the hydro generated to power the operation of the canals.

Ontario Hydro under Chairman Magrath got into the act by hiring Loring C. Christie, formerly of the Department of External Affairs, to prepare a legal case that favoured provincial rights. Christie's arguments were three-fold. One was that the federal government had overstepped its authority over canals by claiming hydro power on all of the Trent Waterway and the Welland Canal. Two was that it was not in the public interest for the federal government to control provincial hydro-electricity, as it would create sectionalism in the country. Finally, water power was a local endeavour and should be controlled by the provinces. The counterargument expressed by the editor of Sifton's *Manitoba Free Press*, John Wesley Dafoe, was that the provinces should not be entitled to hydro power without any expenditure on navigation. This would lead to extensive provincialism and be a threat to Canadian nationalism.

Taschereau and Ferguson met, not so secretly, prior to the federal-provincial conference to plot their strategy. Laurel and Hardy then combined to rally the other Canadian premiers to back their plan for the provinces to retain the rights to hydro power on navigable waterways. They did this in part by supporting motions that would increase transfer payments, then called subsidies, to the have-not provinces. They backed Prime Minister King into a corner that he thought he escaped by referring the issue to the Supreme Court of Canada. Much to King's surprise, the 1929 Supreme Court decision favoured the provinces' rights to hydro power on navigable waterways over the federal government's. King in his memoirs declared that Ferguson was "by nature a skunk,"[73] a popular term for the age.

18
CONCLUSION

"For all sad words of tongue or pen, the saddest are these: It might have been."
—John Greenleaf Whittier

A saying applied to a variety of life's situations is that "timing is everything." Clearly, this applies to the Georgian Bay Ship Canal project. The idea of using the historic route to move prairie grain and minerals east and Nova Scotia coal and eastern manufactures west was an admirable one. If begun early enough in the almost one hundred years of planning and debate, it certainly would have been a success.

Fate conspired against it almost from the beginning. It was up against stiff competition from the Welland–St. Lawrence route, which counted more votes along its shores, especially in Toronto and Southwestern Ontario. Its main proponents were McLeod Stewart and Clifford Sifton, around whom swirled public suspicion. It was an expensive undertaking and came in second in public funding to three transcontinental railways and the Hudson's Bay Railway. It could only operate seven months of the year. It faced competition from already established American shipping routes that were delighted to ship the products of Canada's farms, mines, and manufacturing plants. The canal's death knell was the sound of running water furnishing hydro-electricity to consumers along the St. Lawrence and Ottawa Rivers. The provinces of Ontario and Quebec were able to secure the rights to power along these waterways. In addition, the powerful Quebec Power Trust was against the plan because it would allow the federal government to set rates for the sale of hydro-electricity.

Again, to quote a movie idiom, "build it and they will come" was not in the thinking of decision-makers of the time. There simply was not the collective foresight to realize that this canal would not only improve the movement of goods in Canada but could have been instrumental in creating an all Canadian steel industry, smelting operations, flour mills, etc.

Author Bessel J. Van den Hazel speculates, "What impact would the ship canal have had on the North Bay area if it had been built in 1910? There can be little doubt that the west-east movement of grain, cattle, minerals and lumber would have benefitted greatly from a more direct shipping route to the St. Lawrence River."[74] The opportunities for industrial growth in Northern Ontario would have been greatly enhanced by the canal.

The Georgian Bay Canal continued to capture the imagination of Canadians into the 1970s, when a delegation visited the Minister of Transport to press its construction. The Department of Transport subsequently drafted a memorandum on the subject for the minister. Estimates showed that the toll revenues would be insufficient to cover even the maintenance costs of the canal.

EPILOGUE

Harry Sifton carried on a campaign against Premier Ferguson, who was one of the objectors to the Georgian Bay Canal. In a radio address in July of 1930, he charged Ferguson with being unpatriotic and willing to sell off the resources of Ontario to American interests. In 1930, the battleground for hydro power had shifted to the St. Lawrence Seaway project, in which Harry accused Ferguson and Hydro chairman Magrath of selling Canadian power to the U.S. Harry ran unsuccessfully for a seat in the House of Commons as a Liberal member for North York before devoting himself to a successful land development business. Harry passed away in 1934 due to a cerebral hemorrhage. Winfield had spent most of the post war years in London as a Sifton company promoter and bon vivant. He was a husband for a time to "the best dressed woman in the world," Mrs. Jean Nash. Sadly, he predeceased his brother and father, passing away in 1928 at the age of thirty-eight.

The route has changed over the years in some areas more than others. The Port of Montreal remains the largest inland port in the world, receiving some 2,000 cargo ships in 2019. It also is a popular destination for cruise ships. In 1968, it became the site of Canada's first container terminal. In 1976, the port was moved further east and additional container terminals erected. Subsequently, in the 1990s the Old Port of Montreal was redeveloped as a tourist attraction and now draws six million visitors annually to its mix of theatres, museums, cafés, restaurants, and walking paths.

The provinces eventually won the battle for hydro-electricity, and today there are some fifty hydro dams on the Ottawa River and its tributaries. This power is estimated to be worth one million dollars per day (2019). In 1943, the provinces of Quebec and Ontario reached an agreement to jointly develop and share the power generated. The dams have tamed the river to an extent and created large reservoir lakes behind the dams. The dams altered local ecosystems, interfering with fish and aquatic life by altering the flow characteristics of rivers and creating habitat washout. The dams also altered water quality parameters such as temperature, nutrient concentration, and dissolved oxygen levels. The sixties and seventies were the decades of greatest water pollution in the river due to human and industrial waste. The river is now cleaner due in part to improved municipal waste-water treatment. On a positive note, the large lakes created an environment that cottagers and boaters enjoy. Vital areas of the Ottawa district have come under control of the National Capital Commission, with the goal of protection for future public use and enjoyment.

"Elsewhere in the (Ottawa) valley the provincial governments of Ontario and Quebec have made a start on a few other valuable natural areas. In large result this has been as a direct result of land arrangements following the flooding of shoreline land in the valley by the construction of major dams for hydro-electric development. And it is these same dams that have completely transformed the turbulent river of yesterday into the series of major lakes of today."[75]

The Mattawa River is little changed from earlier days, with the exception of a small power plant and the shoreline development of private camps. In 1970, the provincial government designated a substantial portion of the river as Ontario's first Waterway Park. In 1988, the federal government recognized the historic role of this river in the fur trade and designated it a Canadian Heritage River.

In 1984, a group of citizens from North Bay, under the leadership of North Bay engineer Bill Broughton, proposed a series of locks to allow pleasure boats to navigate between Georgian Bay and Lake Nipissing. This would have been an extension of boating for those who regularly travelled on Georgian Bay. It would also have allowed boaters on Lake Nipissing access to the bay. The opposition to the scheme came from biologists concerned with threats to marine life, canoeists who wanted the French River to continue to be passable by small water craft, and those who feared the reduction in the historical importance of the route. The scheme was never begun.

North Bay today is a transportation hub with three rail lines and Highways 11 and 17 converging on the city. It along with Sudbury are education hubs for Northern Ontario with Canadore College and

Nipissing University operating in the city. It also has a large airport serving as an alternate landing site for Pearson International Airport and an emergency field for NASA's space shuttle. NORAD (North American Aerospace Defense Command) continues in the area, first having built a large underground facility in 1963. In 2006, an above ground-centre that provides air surveillance for Canada and North America was built. The city's population is 51,000, which is small by southern Ontario standards.

The French River was logged extensively between 1875 and 1920, but when the resource was exhausted, there was little left in the way of local employment. Luckily, the river had other assets in the form of game fish. The river is now home to numerous camps or cottages and many fishing lodges.

In 1895–6, the Dokis Restoule Band of Ojibwa heritage moved from the shores of Lake Nipissing to Dokis Indian Reserve #9 on Island Okickendwat on the French River. This island includes the Chaudière rapids, a series of rapids where Lake Nipissing empties into the French River. They built a permanent village and sought to honour their traditional culture. They trapped, fished, raised some animals, worked at lumbering, and sought outside jobs when necessary. With the influx of fishermen, many residents worked as guides during the summer.

In time, the band built a school, commissioned a road into the village, and operated numerous businesses in the area. They remain an honest hard-working people striving to save their language and culture. They erected a large modern band administration building that houses the band offices, medical services, library, etc. There are now approximately 250 band members, not all of whom reside on the reserve. Recently, a joint hydro project was undertaken by the Dokis First Nation and Hydromega Services, built on the reserve. Construction began in 2013, and the hydro-electric plant was operational in 2015. It used a previously constructed man made channel adjacent to the Portage Dam on the French River. Called the Okickendawt Project, its aim was to provide employment for younger members of the reserve.

The St. Lawrence Seaway system for ocean going vessels was not completed until 1959. The Welland Canal portion was finished as early as 1932; however, joint negotiations between Canada and the U.S. over the St. Lawrence Seaway locks continually stalled. All U.S. presidents during the thirties and forties were in favour of the project, but any bill to recommend construction was delayed in Congress. Cities relying on the Erie Canal were concerned about the loss of business, and the cost to taxpayers was considered a hindrance to the project. After prolonged negotiations, Prime Minister Louis St. Laurent told the Americans that if they did not wish to participate in the venture, Canada would build it through exclusively Canadian territory. Congress finally approved participation in 1954.

The Seaway was opened in 1959 by John Diefenbaker, Queen Elizabeth the Second, and Dwight D. Eisenhower. In total, the Canadian government had spent $337 million US on the project and the U.S. $134 million. The system is a series of fifteen locks, thirteen Canadian and two American, between Montreal and Lake Erie. The Canadians also built the Moses-Saunders Power Dam near Cornwall at a cost of $300 million. It boasts an impressive 2,090 megawatts in total capacity of electricity generation.

The Seaway increased in shipping tonnage until 1980 and has been in gradual decline since. At its peak, the Seaway carried 80 million tons of cargo in the form of grain, iron ore, coal, dry bulk, manufactured goods, etc. In 2019, the figure had shrunk to 38,375 million tons. The Seaway does, however, create almost 300,000 jobs between the two countries and generates $45 billion worth of economic activity. It is interesting to note that the maximum draught allowed freighters throughout its entire length is twenty-five feet, a similar number to the proposed Georgian Bay canal.

An unfortunate side effect of allowing ships from foreign ports into the Great Lakes has been the scourge of invasive species that have found their way into the centre of the North American continent. These species came primarily in the ballast tanks of international ships and were dumped into the lakes to allow the ships to load bulk cargo. A host of species including the sea lamprey, zebra mussels and quagga mussels have greatly altered the ecosystems of the lakes and have kept fisheries experts on both sides of the border busy for over half a century. Dan Egan in his book *The Death and Life of the Great Lakes* clearly details this onslaught and the response of scientists and legislators.

AUTHOR'S NOTE

I WAS INTRODUCED TO THE FRENCH RIVER ALMOST THIRTY YEARS AGO, AS MY WIFE'S FAMILY, the Perry clan, has what in Northern Ontario is called a "camp" at Wolseley Bay on the French River. Citizens in my native Muskoka refer to them as cottages or summer residences. Venturing out from the bay by boat, it takes several miles of travel before the camps fade away and rock, water, and pines become the vista. This is the river of the Ojibwa, the Voyageurs, and later the logging companies and the tourists. Since commercial logging ceased about a century ago, the second growth forests are approaching maturity. The area has the feel of wilderness.

The river itself is wide and deep, with the exception of a few propeller-busters that the uninitiated boater may find. Looking down the long channels of the river, it is easy to see why it was considered an ideal location for a canal from Georgian Bay on the Great Lakes to Lake Nipissing and on to the Mattawa River, Ottawa River, and the Port of Montreal.

I have mixed feelings about this long-forgotten project, even though it never came to fruition. It was in its time a good idea to save transportation costs in our vast nation. It could have contributed to interprovincial trade. It was a route secure from interference with our superpower neighbour to the south. If built, it would have brought much more in the way of industry and development to Northern Ontario. It may have helped to create businesses such as shipping, steel manufacturing, and flour milling in the newly minted nation of Canada.

Alternatively, the canal would have created much different water bodies than we "campers" enjoy today. Many of the problems faced during the twentieth century by the companion Ottawa River including loss of wildlife, water pollution and permanent damage due to flooding would have occurred. Many of the invasive species that plague the Great Lakes would have found their way inland sooner than later. Was it a missed opportunity for economic growth in emerging Canada? Certainly, it was. Was the area destined to become an area devoid of much industry but a paradise for lovers of nature and outdoor recreation? It appears so. The bird life pictured below and those that enjoy it are thankful to be without the white noise and unwanted hitch hikers brought by canal freighters.

Grackle

Common Loon

Caspian Tern

Fish Ducks

ACKNOWLEDGEMENTS

I was fortunate to have two amazing photographers provide photos for the book. Terence Hayes, a.k.a. "Slow Ride Photography," is a professional photographer. He is also a kayaker, skier, and outdoorsman. He lives in Sudbury with his wife and two sons.

Jessica McShane is a niece, mother of three, and a photographer of people and places in their natural state. She is a native of Sudbury, where she lives with her husband and children. She can be found at Jessica McShane photography.ca.

In addition to the photographers, I am indebted to numerous family members and institutions for their assistance. My wife Susan was my go-to typist for long and time-consuming quotes. She also has had to put up with my writing and computer use on a daily basis. My mother-in-law, Susie Perry, provided me with unique research materials, as did Mary Waddell, an experienced French River canoeist and historian. My brother-in-law, Robert Perry, encouraged the project and provided helpful insights. His son, Tyler Perry, a French River fishing guide, also contributed to my understanding of the area. Thank you to Sherene Davidson from the Sifton Family Foundation for information on Harry and Winfield Sifton.

Two libraries, the Toronto Reference Library and the West Nipissing Public Library, opened their doors to me during the COVID-19 pandemic and allowed me to complete my research. I am indebted to my editor, Allister Thompson, for his valuable insights and assistance with publishing advice. Finally to the team at Friesen Press and specifically my publishing consultant Stephen Docksteader, thanks for working with me on book four. Hopefully there will be several more to come.

ENDNOTES

1. White, Daryl, "Killing premiers to build a canal: McLeod Stewart and the Montreal, Ottawa and Georgian Bay Canal," *Ontario History*, 99 (2), 146.
2. Smith, Marcus. *The Montreal, Ottawa and Georgian Bay Canal* (Ottawa: Kings Printer, 1895), 6.
3. Perks, Robert W., "The Montreal, Ottawa and Georgian Bay Canal," *Journal of the Royal Society of the Arts*, Vol. 62 No. 3195, 257.
4. *Ibid*, 258
5. Canadian Federation of Boards of Trade. *Georgian Bay Canal A few evidences of Public Opinion*, 12.
6. Morse, Eric. *Fur Trade Routes of Canada Then and Now* (Toronto: University of Toronto Press, 1969), 1.
7. Comrie, Martin C., "The Georgian Bay Ship Canal," *Scottish Geographical Magazine*, 26:1, 1910, 25-6.
8. Legget, Robert. *Ottawa Waterway: Gateway to a Continent*, (Toronto: University of Toronto Press, 1975), 25.
9. Pakkala, H. and M. Hrnjez, *French River: Route to the Past* (Sudbury: The French River Heritage Committee, 1979), 5.
10. Leatherdale, M. *Nipissing from Brulè to Booth*. Second Edition (Victoria: Trafford, 2008), 14.
11. *Ibid*, 21.

12. Legget, *Ottawa Waterway*, 35.
13. *Ibid*, 60.
14. *Ibid*, 50.
15. *Ibid*, 59.
16. Nute, Grace Lee. *The Voyageur* (New York: D. Appleton, 1931), 18.
17. Morse, *Fur Trade Routes of Canada Then and Now*, 5.
18. Legget, *Ottawa Waterway*, 67.
19. Le Belle, W. *Dokis: Since Time Immemorial* (Field, ON: WFL Communications), 13.
20. Jarvis, Eric, "The Georgian Bay Ship Canal: A Study of the Second Canadian, Canal Age 1850-1915," *Ontario History*, 69:2, 1977, 146.
21. Treadwell, C.P. *Arguments in Favour of the Ottawa and Georgian Bay Ship Canal* (Ottawa: Ottawa Citizen, 1856), 5.
22. Harting, T. *French River: Canoeing the River of the Stick Wavers* (Erin, Ontario: Boston Mills Press, 1996), 52.
23. Shanly, W. *Report on The Ottawa and French River Navigation Project*, Submitted to the Legislative Assembly of Canada, July, 1858 (Montreal: John Lovell), 15.
24. Treadwell, C.P., *Arguments in favour of the Ottawa and Georgian Bay Ship Canal*, 40.
25. Jarvis, Eric, "The Georgian Bay Ship Canal," 141.
26. Canadian Parliament, *Debates*, First Parliament, Fourth Session, Vol.6 No. 54 Report on the Commission of Public Works in Canada, Queen's Printer, July 4, 1870.
27. Ritchie, J.A., "Address to the Committee of Railways, Canals and Telegraph Lines", King's Printer, Ottawa, April 26, 1927, 43. "The original gentlemen who were the incorporators of this company were very respectable men. Most of them are now dead, but it might interest you to know who they were, since many of you are very young men and do not know much about those old days. The first name is George Cox. He was once the Mayor of Ottawa.

The next is Mr. McLeod Stewart. He was the chief promoter who had the vision to see this great project and to urge its acceptance upon the people of Canada. He was Mayor of Ottawa also, and is now dead.

The next if Gordon Burleigh Pattee, whith who my friend, Sir Goerge Perley was associated. He and his father were very wll known lumbermen here. He is now deceased.

The next is Henry Kelly Egan. Sir Henry Egan was quite a respectable citizen of Ottawa, lately dead.

John W. McRae; very wll know citizen of Ottawa.

Thomas Birkett; once a member of this House.

Olivier Durocher; once mayor of Ottawa.

Alexander MacLean; once Trade Commissioner in Japan for Canada.

Francis McDougall; now deceased and who was once Mayor of Ottawa, father of Mr. Joseph McDougall, who represented Ottawa for some years in the Local Legislature.

John Charles Rogers; associated with Mr. MacLean as King's Printer.

Dennis Murphy; head of the Ottawa Transportation Company and a very well know citizen of Ottawa.

Charles Berkley Powell. This gentleman is also associated with my friend sir Goerge Perley in the firm of Perley and Pattee. He represented Ottawa in the Local Legislature also.

John E. Askwith; a very well know citizen of Ottawa, and who for many years was our Police Magistrate.

Hon. Francis Clemow; for many years a member of the Senate; now deceased.

Sir James Grant; then a member of Parliament. He was a very eminent physician and attended upon Her Royal Highness, Princess Louise.

Honore Robillard; Member of Parliament at that time.

Thomas Ahearn. Mr. Ahearn is a rather well known citizen, President of the Ottawa Electric Company, and director of the Bank of Montreal and many other great institutions.

George Patrick Brophy; a well-known man at this time.

Alexander Harvey Taylor; a well-known man.

Peter Whelan; a very well-known resident of Ottawa.

Richard Nagle; David McLaren; William Scott; Joseph Kavanagh; Phillip D. Ross, Mr. Ross was one of the chief proprietors of the Ottawa Journal newspaper.

These gentleman were all of Ottawa. Then there were certain people outside of Ottawa. There was Mr. William C. Edwards; then member of Parliament, and later Senator of Rockland.

William T. Hodgins; then member of Parliament, of Hazeldean.

Alexander Fraser of Westmeath; a very well-known lumberman on the Ottawa River.

James Joseph O'Connor of Port Arthur; Arthur Joseph Martin; John Bryson; George H. Macdonnell; Hugh F. McLachlin and Claude McLachlin of Arnprior; and so on.

You see, if this measure was born in sin that it certainly had at the opening very respectable parents."

28. Canadian Federation of Boards of Trade. *The Georgian Bay Canal: A few evidences of Public Opinion*, Montreal, 13.

29. Smith, M., *Report on the Montreal, Ottawa and Georgian Bay Canal*, 23.

30. White, Daryl, "Killing Premiers to Build a Canal," 153.

31. Senate of Canada. *Bill F A Report respecting The Montreal, Ottawa and Georgian Bay Canal Company,* Ottawa, King's Printer, March 13, 1900.

32. Redden, J. *Read! Investigate! Judge and Act! The Welland or Georgian Bay Canal, Which?* Port Arthur Board of Trade, 1913, 3.

33. Department of Public Works Canada, Georgian Bay Ship Canal Report,, 1909, 633-634

NATIVE LEGENDS

"Two Indian legends are associated with this reach of river, which were related to the writer by the Old Chief Pe-ta-wachuan (I hear the rapids far away) known and respected *for many years as "Chief duckies" (Dokis?) at Chaudière portage. Half-way down the reach on the north side, is a great obelisk-like rock that much resembles a huge owl and, in the river, are three small rock islands. Their existence is thus accounted for. Once, long ago, a great hunter of fabulous skill gave chase to a huge owl and three owlets. These he pursued night and day till, in desperation, her little ones becoming exhausted, she threw them into the water, where they instantly became rock brood. Near the foot of the reach is the opening scene of the other tragedy. Here, an ancient land slip has led to the fanciful tale of another great hunter, who was camped with his family nearby when a monster beaver, as shrewd and wicked as he was powerful, stole the hunter's child and retreated to his dam. The infant's piteous cries proclaimed its whereabouts and the frantic father began an attack that breached the dam, as the slide authenticates, but not before the wily beaver managed to escape with the baby and take up a fresh stand behind a curious rock outcrop, some 15 mile up river in the Five-mile rapids. Hither the father pursued and again dislodged the beaver, and this time abandoned the child and beat a hasty retreat across Lake Nipissing and through Trout Lake to a rocky hill between Turtle and Talon lakes. There the beaver was killed with great rejoicing, the whole tribe gathered to feast upon his carcass, but, cut up and in the boiling pot, the tail still splashed the water into foam finally upsetting it, forming Pine lake, which sure enough is 10 feet above all it surrounding neighbours."*

34. Canadian Federations of Boards of Trade. *The Georgian Bay Canal*, 12.
35. Pratte, Andrè. *Wilfrid Laurier* (Toronto: Penguin, 2011), 154.
36. Brown, M.C. and M.E. Prang. *Confederation to 1949: Canadian Historical Documents* (Scarborough: Prentice Hall, 1966), 95.
37. Pratte, Andrè, *Wilfrid Laurier*, 87.
38. Brown, M.C. and M.E. Prang, *Confederation to 1949: Canadian Historical Documents*, 96.
39. White, Daryl, "Killing Premiers to Build a Canal," 163.

40. Redden, J., *Read! Investigate! Judge and Act! The Welland or Georgian Bay Canal, Which?* Introduction.
41. *Ibid*, 10.
42. *Ibid*, 7.
43. *Ibid*, 7.
44. *Ibid*, 7.
45. *Ibid*, 14.
46. Brown, M.C. and M.E. Prang, *Confederation to 1949: Canadian Historical Documents*, 98.
47. Brown, R.C. and Ramsay Cook. *Canada 1896-1921: A Nation Transformed* (Toronto: McClelland & Stewart, 1976), 154.
48. Evans, W. Sandford. *Statistical Examination of Certain General Conditions of Transportation bearing on the Economic Problem of the proposed Georgian Bay Canal, Interim Report*, Ottawa: King's Printer, 1916.
49. *French River Improvement: A Booklet devoted to Economic Transportation* (Toronto: Strathmore Press, 1919), 1.
50. Hall, D.J. *Clifford Sifton Volume 2: A Lonely Eminence 1901-1929* (Vancouver: UBC Press, 1985), 328.
51. *Ibid*, 304.
52. *Ibid*, 306.
53. *Ibid*, 319.
54. *Ibid*, 331.
55. Ferguson Resolution, Ontario Legislature, *Toronto Star*, March 7, 1927. "That the bill proposes to renew a charter to authorize the construction of a canal and the development of water powers on the Ottawa and French rivers, which said charter has already been in existence for thirty-three years without any evidence of progress toward the accomplishment of the projected canal.

That the application to parliament is an effort on the part of the private promoters to secure through the federal parliament the control and ownership of a great and valuable public utility.

That the water powers in the Ottawa River in inter-provincial waters are the joint property of the provinces of Ontario and Quebec, and that the powers in the French river are wholly situated within Ontario and are the property of the province of Ontario and cannot be rightfully legislated upon by the dominion parliament;

That the development of these powers is essential to the industry and the prosperity of the two provinces of Ontario and Quebec, and the only effect of federal legislation purporting to vest these powers in a private company will be to retard development, create litigation and impair the public interest: That the water powers of the Ottawa river, so far as they belong to this province, are an essential part of the public development and distribution of power in Ontario in which the people of this province have already invested upwards of $276,000,000.

That this legislative assembly desires to record its most earnest and emphatic protest against the attempt being made by means of a private bill in the dominion parliament to alienate valuable water powers from the control and ownership of the province, and thereby deprive the people of Ontario of the advantage of one of our greatest natural resources for the benefit and advantage of private promoters.

That the province of Ontario respectfully urges that the rights guaranteed to the province under the federal constitution should be at all times respected by the parliament of Canada.

That this House believes that the occasion calls for a strong and conclusive pronouncement against the proposed legislation as being contrary to the spirit and the terms of confederation and prejudicial to the public interest.

For these and other reasons this House directs that copies of this resolution be forwarded to the Prime Minister of Canada and to the Speakers of the two Houses of Parliament of Canada."

56. *Toronto Star*, October 26, 1926.
57. collections.musee-mccord.qc.ca

58. Westley, M. *Remembrance of Grandeur: The Anglo Protestant Elite of Montreal* (Montreal: Libre Expression, 1990), 202.

59. Vigod, B. *Quebec Before Duplessis: The Political Career of Louis-Alexandre Taschereau.* 122.

60. *Ibid*, 130.

61. *Toronto Star*, March 10, 1927.

62. *Ibid.*

63. Oliver, P.G. *Howard Ferguson: Ontario Tory* (Toronto: University of Toronto Press, 1977), 186.

64. Select Standing Committee on Railways, Canals and Telephone Lines, April 6, 1927, 62.

65. *Toronto Star*, December 4, 1927.

66. *Toronto Star*, April 12, 1927.

67. *Toronto Star*, March 5, 1927.

68. Parliament of Canada. *Debates*, Eleventh Parliament, February 14, 1910, 3531.

69. Select Standing Committee on Railways, Canals and Telephone Lines, April 6, 1927, 36.

70. *Ibid*, 48.

71. See footnote 26 for names and a description of Montreal, Ottawa and Georgian Bay Canal Charter members.

72. Select Standing Committee on Railways, Canals and Telephone Lines, 43.

73. Hall, D.J., *Clifford Sifton: A Lonely Eminence*, 333.

74. Van den Hazel, Bessel, J. *From Dugout to Diesel: Transportation on Lake Nipissing* (Cobalt: Highway Book Shop, 1982).

75. Legget, *Ottawa Waterway*, 214.

LIST OF ILLUSTRATIONS

1. Robert W. Perks. "The Montreal, Ottawa and Georgian Bay Canal," *Journal of the Royal Society of the Arts.* Vol. 62 No. 3195.
2. Jessica McShane Photography.
3. Jessica McShane Photography.
4. Terence Hayes, Slow Ride Photography.
5. Pakkala, H. and M. Hrnjez. French River Route to the Past. Sudbury: The French River Heritage Committee, 1979.
6. Author Photo.
7. Terence Hayes Slow Ride Photography.
8. Map courtesy of David Borys, Langara College, Vancouver, B.C.
9. Department of Public Works Canada. *Georgian Bay Ship Canal Lake Huron to Montreal Twenty Two foot channel project.* Kings Printer: Montreal, 1908.
10. Department of Public Works Canada. *Georgian Bay Ship Canal Vol. 2, Typical views on the projected route.* King's Printer, Montreal, 1908.
11. https://www.thousand wonders.net/Ottawa.
12. Department of Public Works Canada. *Georgian Bay Ship Canal Vol. 2 Typical views on the projected route.* 1908.

13. Department of Public Works Canada. *Georgian Bay Ship Canal Lake Huron to Montreal Twenty Two foot channel project.* 1908.
14. Department of Public Works Canada. *Georgian Bay Ship Canal Vol. 2 Typical views on the projected route.* 1908.
15. Terence Hayes, Slow Ride Photography.
16. Pakkala, H. and M. Hrnjez. *French River Route to the Past.* The French River Heritage Committee.
17. Hopkins, Frances Anne. *Canoe manned by Voyageurs passing a Waterfall*, Library and Archives Canada, C-2771.
18. Hopkins, Frances Anne. *Shooting the Rapids*, Library and Archives Canada, C-2774.
19. St. Laurent, A. *Georgian Bay Ship Canal Plans with Estimates of Cost 1908.* Ottawa: C.H. Parmelee, 1909.
20. Robert W. Perks, "The Montreal, Ottawa and Georgian Bay Canal," *Journal of the Royal Society of the Arts.* Vol. 62 No. 3195.
21. Pakkala, H. and M. Hrnjez. *French River Route to the Past.* The French River Heritage Committee.
22. Jessica McShane Photography.
23. Department of Public Works Canada. *Georgian Bay Ship Canal Lake Huron to Montreal Twenty Two foot channel project.* 1908.
24. Projects.leader.msu.edu.
25. Department of Public Works Canada. *The Georgian Bay Canal, Abstract of leading Facts and Figures from the Report of Government Surveys.* 1909.
26. Department of Public Works Canada. *Georgian Bay Ship Canal Lake Huron to Montreal Twenty Two foot channel project.* 1908.
27. Department of Public Works Canada. *The Georgian Bay Canal, Abstract of leading Facts and Figures from the Report of Government Surveys.* 1909.

28. Ontario Parks, Parks blog, "8 Bucket List Fishing Trips in Northeastern Ontario," Ontario Parks website.
29. Department of Public Works Canada. *Georgian Bay Ship Canal Lake Huron to Montreal Twenty Two foot channel project*, 1908.
30. Terence Hayes Slow Ride Photography.
31. Department of Public Works Canada. *Georgian Bay Ship Canal Vol. 2 Typical views on the projected route.* 1908.
32. IBID
33. Jessica McShane Photography.
34. Department of Public Works Canada. *Georgian Bay Ship Canal Lake Huron to Montreal Twenty Two foot channel project.* 1908.
35. IBID
36. *The Georgian Bay Canal, Abstract of leading Facts and Figures from the Report of Government Surveys. 1909.*
37. IBID
38. IStock photo
39. Wikipedia. After Hafen Von Montreal.
40. Reduto.ca/Great Lakes Ships.
41. Library and Archives Canada ID No. 5033896.
42. McCord Museum, MP-0000.1154.9.
43. Martin, J. "Irrational Exuberance: The Creation of the CNR, 1917-1919," *Canadian Business History*, Joseph L. Rotman School of Management, University of Toronto.
44. Slide from University of Toronto course on Canadian Economic Development, Lecture 6, Confederation and the National Policy.

45. Pakkala, H. and M. Hrnjez. *French River Route to the Past.* The French River Heritage Committee, 1979.

46. Jessica McShane Photography.

47. Library and Archives Canada ID No. 3411051.

48. Hydroquebec.com.

49. ngtimes.ca., Shanahan, D. *Letters of a Local Premier: G. Howard Ferguson.* December 8, 2016

50. Library and Archives Canada ID No. 3623480.

51. Terence Hayes Slow Ride Photography.

52. IBID

53. IBID

54. IBID

55. IBID

56. Jessica McShane Photography.

BIBLIOGRAPHY
BOOKS AND JOURNAL ARTICLES

Brown, R.C. and Ramsay Cook. *Canada 1896-1921: A Nation Transformed.* Toronto: McClelland & Stewart, 1976.

Brown, M.C. and M.E. Prang. *Confederation to 1949: Canadian Historical Documents.* Scarborough: Prentice Hall, 1966.

Campbell, W.A. *The French and Pickerel Rivers: Their History and Their People.* Sudbury: Journal Printing.

Comrie, Martin C. "The Georgian Bay Ship Canal." *Scottish Geographical Magazine,* Vol. 26(1), 1910.

Cook, R. *The Dafoe-Sifton Correspondence 1919-1927.* Altona: Manitoba Record Society, 1966.

Detweiler, D.B. *The Inland Waterways of Canada: Ocean Navigation via St. Lawrence and Welland Route Georgian Bay Route Impractable.* Berlin, Ontario, Great Waterways Union of Canada, 1913.

Dutil, P. and D. Mackenzie. *Canada 1911: The Decisive Election That shaped the Country.* Toronto: Dundurn Press, 2011.

Egan, D. *The Death and Life of the Great Lakes.* New York: W.W.Norton, 2017.

French River Improvement, A booklet devoted to Economic Improvement. Toronto: Strathmore, 1919.

Hall, D.J. Clifford Sifton Volume 2, *A Lonely Eminence, 1901-1929.* Vancouver: UBC Press, 1985.

Harting, T. *French River: Canoeing the River of the Stick Wavers.* Erin, Ontario: Boston Mills Press, 1996.

Holland-Stergar, Patrick. "Depot Harbour: The Rise and Fall of an Ontario Grain Port," M.A. Thesis, University of Western Ontario, London, 2020.

Jarvis, Eric. "The Georgian Bay Ship Canal: A Study of the Second Canadian Canal Age. 1850-1915." *Ontario History* 69:2, 1977.

LaBelle, W. *Dokis: Since Time Immemorial.* Field, Ontario: WFL Communications, 2006.

Latcha, J.A. "Railroads versus Canals." *North American Review*, Vol. 166, No. 495, 1898.

Leatherdale, M. *Nipissing from Brulè to Booth Second Edition.* Victoria: Trafford, 2008.

Legget, R. *The Ottawa Waterway: Gateway to a Continent.* Toronto: University of Toronto Press, 1975.

Macquarrie, K. "Robert Borden and the Election of 1911." *Canadian Journal of Economics and Political Science*, Vol. 25 No. 3, 1959.

Martin, Joe. "Irrational Exuberance: The Creation of the CNR, 1917-1919," *Canadian Business History*. Joseph L. Rotman School of Management, University of Toronto.

Morgan, Robert. "The Georgian Bay Canal." *Canadian Geographical Journal*, 1969.

Morse, E. *Fur Trade Routes of Canada, Then and Now.* Toronto: University of Toronto Press, 1969.

Nöel, Francoise. "Old Home Week Celebrations as Tourism Promotion and Commemoration: North Bay, Ontario, 1925 and 1935," *Urban History Review*, Volume 37(1), 2008.

Nöel, F. *Nipissing Historic Waterway, Wilderness Playground.* Toronto: Dundurn Press, 2005.

Oliver, P. *Public and Private Persons: The Ontario Political Culture, 1914-1934.* Toronto: Clarke Irwin and Company, 1975.

Oliver, P.G. *Howard Ferguson: Ontario Tory.* Toronto: University of Toronto Press, 1977.

Pakkala, H. and Hrnjez, M. *French River: Route to the Past.* Sudbury: The French River Heritage Committee, 1979.

Perks, Robert. "The Montreal, Ottawa and Georgian Bay Canal," *Journal of the Royal Society of the Arts*, Vol. 62, No. 3195, 1914.

Perks, Robert, Obituary. *Journal of the Royal Society of the Arts*, Vol. 83, No. 4282, 1934.

Redden, J. *Read! Investigate! Judge! And Act! The Welland or Georgian Bay Canal Which*, Port Arthur: Port Arthur Board of Trade, 1913.

Rogers, E.S. and D.B. Smith, eds. *Aboriginal Ontario: Historic Perspectives on the First Nations.* Toronto: Dundurn Press. 1994.

Stewart, M. "The Montreal, Ottawa and Georgian Bay Canal Company Letter to the Mayor and City Council of Ottawa." Ottawa: February 8, 1896.

Stewart, M. *Correspondence relating to the Montreal, Ottawa and Georgian Bay Canal.* Ottawa: Thorburn, 1895.

"The Proposed Champlain Ship Canal." *Scientific American*, Volume 30, No. 15, 1874.

The Bankers Magazine and Statistical Register, Vol. 13, Issue 6, 1863.

Treadwell, Charles P. *Arguments in Favor of the Ottawa and Georgian Bay Ship Canal*, Ottawa City: *Ottawa Citizen*, 1856.

Vigod, B. *Quebec Before Duplessis: The Political career of Louis Alexandre Taschereau*. Montreal: McGill Queens University Press, 1986.

Westley, M. *Remembrance of Grandeur: The Anglo Protestant Elite of Montreal*. Montreal: Libre Expression, 1990.

White, Daryl. "Killing premiers to build a canal: McLeod Stewart and the Montreal, Ottawa and Georgian Bay Canal." *Ontario History*, 99(2), 2007.

CANADIAN GOVERNMENT DOCUMENTS

1. Department of Public Works Canada. *Georgian Bay Ship Canal Lake Huron to Montreal Twenty Two foot channel project*, Montreal: Kings Printer, 1908.
2. Department of Public Works Canada. *The Georgian Bay Canal, Abstract of leading Facts and Figures from the Report of Government Surveys*, Ottawa: King's Printer, 1909.

3. Department of Public Works Canada. *Georgian Bay Ship Canal Vol. 2, Typical views on the projected route*, Montreal: King's Printer 1908.

4. Legislative Assembly of Canada. *Debates*, Seventh Parliament, Second Session, Appendix 5. Ottawa: Queen's Printer, 1863.

5. Parliament of Canada. *Debates*, Eleventh Parliament, House of Commons, Ottawa: Kings Printer, Feb. 14, 1910.

6. Parliament of Canada. *Debates*, Eleventh Parliament, King's Printer: Ottawa, March 10, 1911.

7. Parliament of Canada. *Interim Report, The Statistical Examination of Certain General Conditions of Transportation bearing on the Economic Problem of the Proposed Georgian Bay Canal*, Ottawa: King's Printer, 1916.

8. Parliament of Canada. *Debates*, Sixteenth Parliament, First Session, Vol.2 1926-7, Ottawa: Kings Printer, 1927.

9. Parliament of Canada House of Commons Committees. Select Standing Committee on Railways, Canals, and Telephone Lines, Vol. 1. *An Act Respecting the Montreal, Ottawa and Georgian Bay Canal Company*, King's Printer, April 5, 1927.

10. Parliament of Canada. *Sessional Paper No. 8*, First Parliament, First Session, Ottawa: Queen's Printer, 1867.

11. Parliament of Canada. *Debates*, First Parliament, Fourth Session Vol. 6 No. 54, Ottawa: Queen's Printer, July 4, 1870.

12. Parliament of Canada. *Debates*, Third Parliament, Third Session, Ottawa: Queen's Printer, 1876.

13. The Senate of Canada. *Bill F An Act Respecting the Montreal, Ottawa and Georgian Bay Canal Company*, Ottawa: King's Printer, March 13, 1900.

14. The Senate Debates. Tenth Parliament, Fourth Session, *Speech by The Hon. Senator Casgrain on The Georgian Bay Canal*. Ottawa: Kings Printer, 1908.

15. Shanly, W. *Report on the Ottawa and French River Navigation Project*, Submitted to the Legislative Assembly of Canada, Montreal: John Lovell, July, 1858.

16. Smith, M. *Report on the Montreal, Ottawa and Georgian Bay Canal*. Ottawa: Thoburn, 1895.

17. St. Laurent, A., *Georgian Bay Ship Canal Plans with Estimates of Cost 1908*. Ottawa: C.H. Parmelee, 1909.

NEWSPAPER ARTICLES

Huntsville Forester, Dec. 9, 1911

Sudbury Star, Aug. 19, 1952

Toronto Star, June 9, 1894; Feb. 18, 1896; Mar. 9, Oct 6, 7, Nov. 15, 1897; Sept. 8, Mar. 10, June 1, 1898; Jan. 18, May 8, July 21, Nov. 16, 20, 1899; April 23, 1901; Mar. 28, June 11, 1905; Mar. 14, May 17, June 2, Sept. 8, Dec. 14, 1906; Jan. 17, June 11, July 23, 1907; July 4, Sept. 5, 1908; Jan. 25, May 12, Oct. 9, 1909; Jan. 25, Mar. 3, May 27, 1910; Jan. 22, Feb. 8, 15, 24, Mar. 27, Dec. 23, 1912; Feb. 4, Aug. 6, 1913; Feb. 24, 27, Mar. 12, 30, April 9, Aug. 23, 1914; April 3, 1919; Mar. 15, 1921, Mar. 2, Oct. 12. 1922; Jan. 16, May 1, July 30, Oct. 12, 1925; April 28, May 1, Sept. 7, Oct. 4, 6, 26, 1926; Feb. 5, 28, Mar. 1, 2, 4, 5, 8, 10, 12, 15, 16, 19, 21, 22, 26, 29, April 2, 5, 6, 7, 11, 14, 23, 30, Oct. 5, 1927; Jan. 12, 1928; Feb. 6, 1929; April 21, 1930.

CPSIA information can be obtained
at www.ICGtesting.com
Printed in the USA
BVHW062205291021
619824BV00004B/79